CONCEPTS OF SCIENCE EDUCATION

A Philosophical Analysis

Michael Martin
Boston University

UNIVERSITY PRESS OF AMERICA

LANHAM • NEW YORK • LONDON

111926

Excerpts from *Biological Science: An Inquiry into Life,* Harcourt Brace Jovanovich, Inc., New York, reprinted by permission of the Biological Science Curriculum Study.

To my mother

PREFACE

As pursued by professional philosophers, philosophy of science is a technical and demanding study. It is small wonder that science educators are, by and large, not centrally concerned with philosophy of science and somewhat unaware of the philosophical literature on science. Nevertheless, this lack of awareness is unfortunate. As I will show in this book, philosophy of science has a great deal to offer science educators.

Conversely, philosophers of science are virtually oblivious to the concerns of science educators. They may, of course, have some vague idea that philosophy of science is relevant to science education. But in general they have made very little effort indeed to show this relevance. Had they done so, they might have come to appreciate (as I have in writing *Concepts of Science Education*) the practical uses of philosophy of science for education, and perhaps have come to understand philosophy of science itself in greater depth.

The purpose of this book, then, is to bridge the gulf that now exists between philosophy of science and science education. I have tried to make philosophy of science meaningful and relevant to science educators without watering it down or distorting it. In so doing, I also hope I have made philosophy of science more alive to philosophy students by showing them what philosophy of science has to say in a practical area. Thus this book aims at being an introduction to philosophy of science for science educators — the people studying science education in schools of education, the people already teaching science in grade schools or high schools, the people designing science curricula, the people writing science textbooks, and the people doing research in science education — as well as for typical philosophy students — the people studying philosophy in liberal arts colleges. Moreover, for all readers *Concepts of Science Education* aims at being an introduction to science education in a way that is unique — via philosophy in general and philosophy of science in particular.

The difficult task of showing the relevance of philosophy of science to science education was first set for me by the editor of this series, Israel Scheffler, and I am grateful to him for the challenge, as well as for his help and encouragement as I tried to carry it out. James Rutherford, the science education consultant for this book, has also been very helpful. His constant question, "Would this be meaningful to science educators?" has forced me to eliminate philosophical jargon, clarify my prose, and search for interesting examples. That *Concepts of Science Education* may still have faults is no fault of theirs.

Many other people have helped me and they should be mentioned. My classes in philosophy of science and science education at Harvard University Extension have proved an invaluable sounding board for my ideas. During the year 1969 – 1970 I was a Fellow at the Center for the Advanced

Study of Theoretical Psychology, University of Alberta. I am indebted to the Director, Joseph Royce, and members of the Center for providing me time, facilities, and the intellectual stimulation necessary for completing the manuscript. My mother, Ruth Martin, to whom this book is dedicated, did much of the typing of early drafts of the manuscript and I appreciate her efforts. I thank the secretarial staff of the Boston University Philosophy Department, who worked long and hard to produce the final copy of the manuscript. Finally my wife, Jane Martin, typed, edited, criticized, encouraged, and lent moral support. Without her help this book would never have been completed.

Michael Martin

TABLE OF CONTENTS

INTRODUCTION

What relevance – if any – does philosophy of science have for science education? Unfortunately, this question has been largely unexplored. To be sure, a great deal has been written on the philosophy of science; perhaps even more has been written on science education. However, surprisingly little has been written on the relation between the two areas. One aim of this book is to help remedy this situation. Although this book is primarily intended to be an introduction to the philosophy of science, it can also be used as an introduction to science education – at least in its theoretical and philosophical dimensions. One of the major theses of this book is that philosophy of science has great relevance for science education. This relevance may best be seen by considering some typical problems faced by science educators: science teachers, curriculum planners, researchers in science education, and science textbook writers.

Consider a science teacher whom we will call Miss Evans. Miss Evans is interested in teaching science with the latest methods and insights. She has been particularly impressed by the recent emphasis on scientific inquiry in teaching science. Her reading of curriculum theorists like Joseph Schwab[1] has convinced her that science should be taught as inquiry, and she plans to put her ideas into practice in her classroom. What relevance does philosophy of science have for the concerns of Miss Evans?

Philosophers of science have analyzed scientific inquiry. They have proposed different accounts of this inquiry, separated its different aspects, and analyzed its logical structure. Some knowledge of philosophy of science would undoubtedly help Miss Evans understand scientific inquiry and the problems connected with it. Furthermore, knowledge of philosophy of science might give her new ideas and insights about teaching science as inquiry, for philosophy of science can be pedagogically suggestive. We will see in detail in Chapter One how philosophy of science can clarify the thinking of science educators about scientific inquiry and also suggest new thinking. But the clarificatory and suggestive powers of philosophy of science have a wider range than is revealed in this example.

Consider a biological science teacher, Mr. Smith. Today's discussion in Mr. Smith's class concentrated for a few moments on the extinction of the dinosaurs. The students asked why these reptiles disappeared after a reign of millions of years. Mr. Smith, as well as the textbook he is using, pointed out that, although several theories of the extinction of dinosaurs have been given, no account is completely adequate, that the question has yet to be answered satisfactorily. The discussion then moved on to a new topic.

A classroom situation of this sort — in which a question is posed by the students which cannot be answered by the teacher, not because of his incompetence, but because of the state of the field — is familiar to science teachers. Yet it raises certain philosophical issues which are important in their own right, and which may have great pedagogical value. What the students wanted and what Mr. Smith was unable to give was a scientific explanation of a certain fact, the disappearance of dinosaurs. An obvious question arises: What would be a satisfactory explanation of the dinosaurs' disappearance? What standards would such an explanation have to meet in order to be satisfactory? Clearly, unless we have some idea of these standards, we have no good reason for rejecting explanations already offered, nor will we know when an adequate explanation comes along. This question in turn raises certain general questions: Do all scientific explanations have to meet the same standards? If not, why not? If so, what are these standards? These general questions about explanations are questions that philosophers of science are concerned with. But they are also of interest to science educators, and to Mr. Smith in particular.

For suppose that philosophical inquiry reveals that scientific explanation must meet certain standards, and that, in the case of the theories so far offered to explain the disappearance of the dinosaurs, these standards have not been met. Such knowledge might certainly have helped Mr. Smith. For example, in today's discussion he might have told his class (or led them to discover) what standards the existing theories of the disappearance have not met. He might have brought this out by pointing out differences between these theories and acceptable theories in other areas. This procedure would be an improvement over what did happen. The class learned that existing accounts were inadequate; they did not learn why they were inadequate, they did not learn to see differences between the acceptability of theories in different domains. We shall discuss scientific explanation in detail in Chapter Two and show the relevance of philosophical analysis of explanation to science education.

Consider another example. Mr. Brown, who is chairman of the science department in the local high school, is interested in developing a new science curriculum for his school. He has been impressed by recent concern with the teaching of the structure of a subject and he wishes to build the curriculum he is developing around certain basic scientific procedures and concepts.[2] His reading of Bruner in particular has convinced him of the importance and centrality of operational definitions in science. He begins to develop a physics curriculum in which operational definitions of key physical terms play a crucial role. How could Mr. Brown be helped by philosophy of science?

First, a study of philosophy of science would help Mr. Brown understand what an operational definition is. Philosophers of science have analyzed definitions in detail. Secondly, philosophy of science would help him understand some of the limitations of and problems connected with operational definitions. Philosophers of science have evaluated the use of operational definitions in science. Some knowledge of philosophy of science is surely crucial in developing a science education program concerned with operational definitions, and in appraising the importance of operational definitions in a science curriculum.

A discussion of operational definitions in science in turn raises crucial philosophical and pedagogical questions about scientific definitions in general: What is a definition? Are there different types of definitions? What is the relevance of scientific definitions for science education? These questions, as well as questions about operational definitions in science and the teaching of science, will be taken up in Chapter Three.

Mr. Soames, a science textbook writer, has recently turned his attention to social studies texts and is in the process of writing a book on social studies with a strong emphasis on social science findings and methods. One social science method which he wants to discuss in his forthcoming book is participant observation, which is used extensively in anthropology. Can philosophy of science be relevant for Mr. Soames?

Philosophical analysis can help Mr. Soames understand what participant observation is and what the various problems and limitations of such a method are. Furthermore, a consideration of observation in science — participant observation or otherwise — raises crucial questions about the objectivity of science. It might well be asked whether all observation is biased by the background beliefs and attitudes of the observer. Can there be such a thing as observation that is not tainted by theory? In any case, some knowledge of the philosophical problems connected with observation could surely enrich Mr. Soames' writing. We shall consider some of the philosophical issues connected with observation in general and participant observation in particular, and their relation to science education, in Chapter Four.

Mr. Goodwin is an educational researcher who is interested in evaluating the effectiveness of a new science curriculum project. His major concern is whether this project has been effective in achieving certain goals. Mr. Goodwin has been influenced by the recent emphasis on behavioral goals, and plans to use such an approach in his research.[3] Does philosophy of science have relevance for Mr. Goodwin's concerns?

It does. Philosophy of science can shed light on the methodological motivation for the behavioral goals approach, and on its problems and limitations. In Chapter Five an analysis and critique of the behavioral goal approach is given.

Mr. Brown, the curriculum planner, faces a different problem in connection with the goals of science education. He wants the science education program in his school to aim at certain goals. He feels strongly that some goals are appropriate for science education and some are not. Yet he finds it difficult to justify his choice of goals. Moreover, he has the disturbing thought that perhaps all goals of science education are arbitrary, and that no one set is any better than any other. What relevance does philosophy of science have for Mr. Brown's worry?

Philosophical considerations may well give Mr. Brown good reason to pick one set of goals rather than another. In Chapter Five we will consider whether certain goals of science education can be rationally justified and what these goals might be.

The major concerns of this book should by now be clear. Chapters One through Four will be a philosophical examination of some of the central ideas in science: inquiry, explanation, definition, and observation. These will be analyzed and their relevance for science education considered. Chapter Five will consider the goals of science both methodologically and normatively. The methodological issue of behavioral goals and the normative issue of which goals are rationally justified will be considered.

Readers unacquainted with recent developments in philosophy of science may be surprised at the philosophical orientation taken in this book. Metaphysical speculation about science is almost totally absent. Thus there is little philosophizing about science in the traditional manner. The major emphasis is on analysis and clarification, but this emphasis does not mean that the author is opposed to philosophical speculation about science — what has been called speculative philosophy of science. On the contrary, philosophical speculation about the findings of science may be a worthy intellectual pursuit in its own right and may in fact have heuristic value for science and science education. However, it is the author's belief that analytic philosophy of science — the kind found in this book — has an extremely useful role to play both in understanding science and in illuminating science education. Analytic philosophy of science is therefore emphasized here, although the relevance of traditional speculative philosophy of science to science and science education is not denied.[4]

But analysis and clarification are not our entire concern. In each of the first four chapters, we shall make certain educational recommendations that are suggested in part by the results of our analysis. We shall show that analytic philosophy of science has suggestive as well as clarificatory power for science education. In the last chapter we will deal explicitly with normative issues, that

is, determining which goals of science education are rationally justified. Although some of Chapter Five consists of analysis and criticism, this will be aimed in part at defending our normative recommendations about the goals of science education.[5]

One of the major theses of this book is that the study of philosophy of science can help science educators in their thinking
about science education and in their educational practice. This is
not the same as saying that the study of philosophy of science will
help science students, the people science educators have in their
classes. To be sure, if the study of philosophy of science helps science educators in their thinking and practice, it will indirectly
help science students. But it does not follow from this that science
students should study philosophy of science. In a similar manner,
the study of child psychology may help science educators in their
thinking about and practice in science education and thus indirectly help science students. From this it does not follow that science
students should study child psychology. While there is some evidence to suggest that it would be useful for science students to
study philosophy of science,[6] this is not one of the major theses of
this book. Whatever the reader may think about science students'
studying philosophy of science, we hope to convince him that at
least science educators would be well advised to do so.[7]

CHAPTER ONE

•

Scientific Inquiry

•

One of the most interesting chapters in the history of science is that of the overthrow of the theory of spontaneous generation One part of this story is the work of Francisco Redi. During Redi's time many people supposed that the maggots which appeared on meat were generated spontaneously. Redi suggested instead that the maggots came from eggs laid on the meat by flies. He then went on to test his hypothesis by some simple experiments.[1]

The case of Redi is typical of scientific inquiry. In it we can discern two aspects: first, the generation of Redi's hypothesis; second, the testing of his hypothesis. The first aspect will be referred to here as the *context of generation* and the second as the *context of testing*.[2] Both contexts raise philosophical problems and have important implications for science education.

In the context of generation, for example, one might well wonder whether there is any connection between the way a hypothesis like Redi's is generated and the truth of that hypothesis. One might wonder whether any reliable way of generating hypotheses like Redi's exists, whether there is some procedure of hypothesis generation one can follow. Perhaps hypothesis generation is a completely irrational affair—a matter of luck or inspiration. On the other hand, perhaps there is a definite logic of scientific discovery.

If there is a logic of scientific discovery, what relevance does this have for teaching science? In particular, what should science teachers be doing in their teaching that they are not doing now? Could insights derived from philosophical investigations of the context of generation be helpful in utilizing certain apparatus in science classrooms?

Different questions arise in relation to testing. For example, what arguments are used in testing a hypothesis like Redi's? Are there different theories or philosophical accounts of these testing procedures? If so, which is correct? Do science educators, e.g., science textbook writers, have an adequate view of the philosophy of hypothesis testing? In particular, is the Redi example as discussed in science textbooks well handled? What implications for textbook writing and classroom procedure do different philosophical theories of scientific theory testing have?

These are some of the questions that will be taken up in the second part of this chapter. In the first part we will discuss questions about the context of generation. In connection with both contexts, we will ask what relevance a given issue has for science education.

TEACHING SCIENCE AS INQUIRY

Before discussing in detail the context of generation and the context of testing, we should consider the relationship of our discussion to the recent emphasis on teaching science as inquiry. We noted in the Introduction that teachers like Miss Evans are concerned with teaching science as inquiry, and we suggested that philosophical analysis has relevance to this concern. This suggestion needs further argument.

What does teaching science as inquiry involve? What relevance could philosophical analysis have for it?

TO TEACH THAT SCIENCE IS TENTATIVE AND FLEXIBLE

Often when science educators argue that science should be taught as inquiry, they are arguing that students should learn to view science in a particular way, that they should see scientific findings as tentative, flexible — not God-given, but subject to revision and reformulation.[3] This way of viewing science is in contrast with what is alleged to be the traditional way of viewing science, in which scientific findings are taken for granted, considered to be unchanging and eternal.

If this is what teaching science as inquiry involves, a philosophical analysis of scientific inquiry is surely important. For as we shall see in detail later, an examination of the philosophy of theory testing gives a clear rationale for looking at science as something flexible and subject to change. An examination of the context of testing indicates quite clearly that the findings of science are never certain. A teacher who knew something about the philosophy of hypothesis testing would indeed have a clear insight into the tentative character of science.

TO TEACH THE WAY IN WHICH SCIENCE ARRIVES AT ITS FINDINGS

What has also been meant by teaching science as inquiry is an approach to teaching science which emphasizes the *way* in which science arrives at its findings, rather than the findings themselves. The phrase "the way science arrives at its findings" is ambiguous, however. It can refer to the context of generation or to both the context of generation and the context of testing. Let us understand the phrase in the latter sense. In any case, this approach to teaching science is in contrast with an approach which emphasizes the findings of science, rather than the way scientists arrive at their findings.

Three different things might be meant by putting emphasis on the way science arrives at its findings. First, one might mean something quite intellectual. The emphasis could be on students' *learning about* the context of generation and the context of testing. For example, there might be strong emphasis on studying cases in the history of science in order to learn about how hypotheses are generated and tested. Secondly, the emphasis might be on something quite practical and nonintellectual, such as students' *learning how* to generate and test hypotheses. Students could be given extensive practice in hypothesis generation and testing. Thirdly, the emphasis might be on both the intellectual and the practical aspects. For example, stress could be placed on both *understanding* the way science arrives at its findings by studying cases from the history of science and *learning how* to generate and test hypotheses by actual practice.

All of these interpretations require philosophical discussion of scientific inquiry. For the way science arrives at its findings involves the generation and testing of certain hypotheses—precisely the two aspects of scientific inquiry we are about to discuss. A teacher with a clear understanding of the context of generation and the context of testing will have a clear understanding of the way science arrives at its findings. This will be useful regardless of whether the stress is on having students acquire intellectual knowledge, or practical skill, or both, about the way science arrives at its findings.

TO TEACH SCIENCE BY THE DISCOVERY METHOD

So far we have been concerned with what has been meant by teaching science as inquiry, and with *what* is taught, that is, (1) how science is changeable and flexible, and (2) how science arrives at its findings. However, sometimes when educators speak of teaching science as inquiry they are referring to the *way* science is taught, rather than to what is taught.[4]

Considered in this way, teaching science as inquiry involves what is called the *discovery method*. This method is to be contrasted with other methods of learning, e.g., by being told, reading books, attending lectures, serving as an apprentice. With the discovery method, students are supposed to figure things out for themselves and arrive on their own at scientific findings which they might, with a more traditional approach, arrive at by reading, hearing a lecture, and so on.

Advocates of this approach claim it has great advantages. Retention and transfer value are supposed to be greater if knowledge is acquired in this way. Further, it is claimed that students learn how to inquire by being taught by the discovery method. At the same time, strong criticisms and serious reservations have been expressed about this approach to teaching.[5]

What is important to notice here is that people who advocate teaching science as inquiry in either of the first two senses need not be committed to the discovery method. One might well advocate that students should learn that science is flexible and changing, or that the emphasis in science education should be on how science proceeds rather than on its findings, without advocating the discovery method of achieving this learning.

Whatever the problems of the discovery method may be, philosophical analysis of scientific inquiry could be useful for advocates of the discovery approach. One of the things learning by discovery is supposed to teach is how to inquire. Thus advocates of the discovery method have seemed to suppose that students taught by the discovery method will know how to do scientific inquiry better than students who are not taught by this method. Furthermore, some advocates of the discovery method have thought that when students figure things out for themselves—as they are supposed to do in the discovery method—they will be engaging in scientific inquiry. However, in order to evaluate these claims, one must have a clear idea of what scientific inquiry is.

It is hoped that this discussion of what teaching science as inquiry might mean has convinced the reader that philosophical analysis is relevant to teaching science as inquiry. Let us now consider scientific inquiry in detail, and some of its implications for science education.

THE PROCESSES BY WHICH HYPOTHESES ARE GENERATED

Scientists come to new hypotheses by many different processes. Some hypotheses occur to scientists in their waking hours after much conscious thought and deliberation, some occur in their dreams, some occur suddenly in a flash of insight, and some come slowly and painfully. Often it seems that it is only the sudden flashes or dreams resulting in true hypotheses that students remember. As every student of science knows, Archimedes grasped the principle of displacement in a flash while stepping into his bath, and Kekulé came upon the chemical makeup of benzene in a dream before a fire. What is often forgotten is that not all hypotheses occur to scientists in such spectacular ways, and some of those that do are false. In any case, it seems clear that the process by which a hypothesis comes to a scientist is logically irrelevant to the truth or falsity of the hypothesis. Thus, a hypothesis that comes to a scientist from hitting his head against the wall or in any other bizarre way may still be true, while a hypothesis that comes to a scientist as a result of careful study and chains of valid arguments may be false.

This does not mean, however, that all processes by which hypotheses are generated are equally good. Some processes may produce more true hypotheses than others, some processes may produce more plausible hypotheses than others, some processes may produce more easily testable hypotheses than others. Thus, to admit that the process by which a new hypothesis is generated does not determine with complete certainty the truth or falsity of the hypothesis is quite compatible with saying that there are some processes which are more successful in generating new hypotheses with desirable properties (truth, plausibility, testability, and so on) than others. Moreover, even if all processes are equally successful in generating hypotheses with these desirable properties, some processes may still be better in other respects. For instance, some processes may be less painful or less expensive.

It is important to realize that, even if a process were known to produce new hypotheses with some desirable property at a higher relative frequency than other processes, it would not follow that one could deliberately adopt such a process. First of all, certain processes may be impossible for most people to adopt. Suppose it were found that concentrated study of a certain kind produced new hypotheses with some desirable property at a high frequency. It might be impossible for most people to study in this way and thus to utilize this process of generating new hypotheses. Moreover, some processes may be impossible for anyone to adopt. Suppose people in epileptic seizures came up with true hypotheses at a higher relative frequency than people who were not having epileptic seizures. Unless one knew how to bring about these seizures it would be impossible to adopt this process of generating hypotheses deliberately. One would have to let nature take its course.

The previous paragraph suggests two quite different types of processes by which new hypotheses are generated: processes that can be deliberately adopted in order to generate hypotheses and processes that cannot be. The processes that can be deliberately followed in order to generate hypotheses will be referred to as *methods* for generating hypotheses, and processes that cannot be deliberately followed will be referred to as *nonmethods*. Methods would include bizarre techniques such as hitting one's head against the wall, as well as standard techniques such as brainstorming. Having dreams and epileptic seizures would qualify as methods only insofar as they could be brought about deliberately. But what can be brought about deliberately at any time is in part a function of the state of technology at the time. Thus there can be no fixed line between methods and nonmethods for generating hypotheses; the very thing that at one time could not be brought about deliberately could perhaps be brought about deliberately at a later time.

Furthermore, just because a process of generating new hypotheses is a method does not mean that whenever a person is doing what is specified by the method, the person is *using* the method. For instance, suppose concentrated study on a particular problem is a method for generating new hypotheses about the problem. Nonetheless, someone who is studying a problem may not be studying it in order to generate new hypotheses; he may only want to learn about the problem. We must distinguish, then, between someone who is simply doing what is specified by the method and someone who is using the method. Nonetheless, people who are doing what is specified by the method and not using the method may still come up with new hypotheses as a result of what they are doing. For example, the person who studies a problem because he wants to learn about the problem may come up with new ideas about how to solve the problem.

STAGES OF HYPOTHESIS GENERATION

Some authors have suggested that the generation of new hypotheses comes in stages. One of the best-known attempts to formulate stages of hypothesis generation is that of Graham Wallas.[6] Wallas called the stages *preparation, incubation*, and *illumination*. In the preparation stage the would-be hypothesis generator prepares himself by diligent study of the area in which a new hypothesis is needed. He masters the data and theories of the area and perhaps of related areas from which fruitful analogies could be drawn. In the incubation stage the hypothesis generator refrains from his intellectual efforts. He rests or turns to other work. The unconscious takes over and works on the material assimilated by conscious effort in the preparation stage. In the illumination stage the new hypothesis comes to consciousness, either when the hypothesis generator returns to work on his problem or before he returns. The new hypothesis comes effortlessly and quickly in a flash or click of insight.

All sorts of questions can be raised about the truth of Wallas' account. For instance, do these three stages — especially the last two — accurately describe the stages that most scientists go through in arriving at new hypotheses? Is it really true that most people in science arrive at new hypotheses in a flash, that new hypotheses come quickly and effortlessly, or would it be more accurate to say that for most scientists new hypotheses arise slowly and with great mental struggle? Again, is it true that scientists usually arrive at new hypotheses after they have been away from their problem, after moments of relaxation or diversion, rather than while working on the problem without a break? Although Wallas' account certainly describes the stages *some* scientists go through in arriving at new hypotheses, e.g., Archimedes and Kekulé, whether his account has wide application is an open question.

However, the accuracy of Wallas' account is not as important as other things about it. First, even if Wallas' account is accurate, it does not follow that going through these three stages is the best process by which hypotheses are generated. Perhaps going through these three stages does not generate as many hypotheses with some desirable property, e.g., truth, as some other process. Conversely, Wallas' three stages might be the best process by which hypotheses are generated, without being an accurate description of how most people arrive at hypotheses.

Secondly, we must ask in what sense these three stages constitute a *method* for generating hypotheses. It seems quite clear that most of stage two, incubation, and all of stage three, illumination, cannot be deliberately brought about, at least in the present state of our technology; hence they cannot be part of a method of generating hypotheses. To be sure, a method of generating hypotheses can be abstracted from these three stages, but it will not include every aspect of each stage. One *can* prepare oneself by study and one *can* later refrain from this intellectual effort. We will call a process some but not all of whose sequences of conditions can be deliberately followed a *partial method*. Wallas' three stages then can be viewed as a partial method of generating hypotheses, since some of the sequences can be deliberately followed and some cannot.

Some philosophers have argued that the generation of new hypotheses — unlike the testing of new hypotheses — is not rational. Reason comes in the context of testing but not in the context of generation. However, it is not clear what this thesis amounts to.

One meaning might be that there is no *mechanical method* for generating new hypotheses with some desirable property.[7] By a mechanical method is meant a procedure that leads in a finite number of predetermined and routinely followable steps to the answer, i.e., to a set of hypotheses with the desired characteristics, such as truth. It is true that there is no known mechanical method for generating new hypotheses of this kind, but this does not mean that the generation of hypotheses is not rational. The existence of a mechanical method is not coextensive with rational procedure. There is no mechanical method for generating proofs in certain parts of logic or mathematics, but this does not mean that the generation of proofs in these parts of logic and mathematics is nonrational. There may be good strategies or heuristic procedures for arriving at proofs even when there is no mechanical method for generating such proofs.

Saying that the generation of hypotheses is not rational might also mean that one process by which hypotheses are generated is no better than another, that one could not have a good reason for preferring one process to another so any choice is arbitrary. If this is what is meant, the contention is probably false. Hitting one's head against a wall probably does not generate new hypotheses with some desirable property as frequently as other ways. Moreover, as we have already seen, even if hitting one's head against a wall were just as successful as other ways at generating hypotheses with some desirable property, this would not mean that other processes would be no better. Some other ways would be preferable on the grounds of utility: they would be far less painful.

The contention that the generation of hypotheses is not rational may simply amount to the claim that no method for generating new hypotheses with some desirable property is as successful as some nonmethods. The claim may be that the generation of hypotheses is better left to happenstance than to purposeful activity, that successful hypothesis generation is not rational because it is not controllable. This certainly is dubious, for there seem to be methods of generating new hypotheses that are as likely to result in hypotheses with certain desirable properties as any uncontrollable process. For instance, it would seem that the method that can be abstracted from Wallas' three stages, namely preparing oneself by study and then refraining from this, although it may not be the best method of generating hypotheses, is likely to be better than any process that is completely beyond control.

The contention that generating new hypotheses is not rational may be interpreted in a slightly different way. The contention may be that some methods of generating new hypotheses *combined with* nonmethodological elements are more successful in generating new hypotheses with certain desirable properties than any method alone. Put in different terms, the contention may be that some partial methods are more successful in generating hypotheses with some desirable property than any method, that there is an element of the nonrational, the nonmethodological, in any successful way of generating hypotheses.

This last contention may well be true; indeed, it seems quite plausible. However, even if it is true, it does not follow that the nonmethodological elements are forever beyond control. Consider Wallas' stages of incubation and illumination. These may now be beyond conscious direction, but future developments in psychology and physiology may allow us to control these factors to a certain extent. Thus, some of the nonrational (nonmethodological) elements in the successful processes of hypothesis generation may, at least in time, be subject to some control. Moreover, successful partial methods may closely approximate some methods of generating hypotheses; there may be little left to happenstance.

In sum, rationality in the context of generation seems to come down to two questions: (1) Are there certain processes by which hypotheses are generated that are more successful than other processes, i.e., produce a higher relative frequency of hypotheses with certain desirable properties than other processes? The answer is that there very likely are. (2) Are these successful processes methods? The evidence is unclear but the tentative answer is that the most successful processes are probably partial methods. There may, however, be partial methods that approximate closely to methods, and some of the nonmethodological elements of the partial methods may in time be subject to control.

What exactly the successful partial methods are is not known. However, that Wallas' three stages are the most successful partial method is a conjecture worth further investigation. The final answer must, of course, be determined by further empirical research. This discussion may be considered only as a way of clearing the ground and making suggestions for further research.

The view suggested here that there may be rational processes by which hypotheses are generated, i.e., processes largely under conscious control, that are more successful in the generation of new hypotheses with certain desirable properties than others, should be distinguished sharply from another thesis, that there is a *logic* of scientific discovery. For what has been meant by a logic of discovery is not a way of generating new hypotheses. Rather, recent discussion of a logic of discovery has centered on the reasons that can be given for supposing that a certain *kind* of hypothesis is true after the hypothesis is generated but before any test has been made of the hypothesis.[8] Thus it has been argued that Kepler had good reason to suppose that Jupiter's orbit was noncircular before any particular noncircular hypothesis was tested. Consider:

(1) Mars' orbit was noncircular,

and

(2) Mars was a typical planet.

According to some philosophers of science, (1) and (2) gave Kepler good reason for supposing that:

(3) Jupiter's orbit was noncircular

before (3) was tested.

(The argument put in a more formal way is this:

Mars' orbit was noncircular.
Mars was a typical planet.
Jupiter was a planet.
If x_1 is a typical member of B, then if x_1 has property P and x_2 is a member of B, then x_2 frequently has P.
(Probably) _____
Jupiter's orbit was noncircular.)

Thus the existence of a logic of discovery does not seem relevant to what is at issue here, rationality in the context of generation. However, the context of generation is not completely irrelevant to a logic of discovery. One reason for supposing that a hypothesis is true before testing the hypothesis is the way in which it was generated. Suppose Kepler knew:

(1′) The hypothesis that Jupiter's orbit was noncircular was generated by process P.
(2′) Process P results in more true hypotheses than not.

Then (1′) and (2′) could have also been used by Kepler to support (3). To be sure, whether there is any process by which hypotheses are generated that results in more true hypotheses than not is not known. However, there does not seem to be any *a priori* reason why there could not be such a process.

SCIENCE EDUCATION: HYPOTHESIS MACHINES AND THEORY BOXES

Some science educators have developed "black box" apparatus — variously called hypothesis machines and theory boxes — as classroom tools. An apparatus is presented to the class, but its inner mechanism or working is either completely or partially hidden. Something is put in the apparatus (the input) and something comes out of it (the output). Students are then asked to guess what hidden mechanism produced the output, given the input.

For example, in one apparatus clear water is poured in at the top of the box and colored water comes out at the bottom.[9] Students are asked what hidden mechanism produces the colored water from the clear water. In another apparatus is a series of parallel transparent tubes.[10] However, the middle section of the tubes is covered. When marbles are rolled down a tube they disappear behind the covered section; sometimes they appear again in the same tube on the other side of the covered section, sometimes they appear in a different tube, and sometimes they do not appear at all. Students are then asked to guess what hidden mechanism in the covered section caused the observed behavior of the marbles.

The inventor of the first apparatus argues that students should have the opportunity "to become familiar with the role and functioning of theories,"[11] and he seems to believe that his apparatus offers such opportunity. The inventors of the second apparatus argue that this apparatus teaches "indirect observation."[12] However, it is equally plausible to suppose that these apparatus teach students how to generate hypotheses. Indeed, hypothesis generation seems to be what is stressed in the actual use made of the apparatus by these inventors.

With the first apparatus, "each student is asked to draw — without collaboration — the mechanism which *he thinks* is contained in the box. After 5 or 10 minutes, the teacher selects a half dozen students to diagram on the chalkboard what they have on their papers."[13] The inventors of the second apparatus say that "students are asked once again for hypotheses. The class is provided with actual size drawings of the machine so that they can test their hypothesis by drawing what they think is in the target area (the covered part of the tubes)."[14] The word "test" is misleading. The drawing that is asked for is not a *test* of the hypothesis, but a graphic way of representing the hypothesis which the student has generated.

In any case, although these apparatus are used to teach hypothesis generation, they have yet to be used to their fullest capacity. First, if there is anything at all to Wallas' partial method, students should be well prepared before they attempt to generate hypotheses. These students are not well prepared; they are often asked to generate hypotheses immediately, and without any relation to knowledge they have acquired elsewhere. A more fruitful approach might be to have students do some preparation after the problem is posed and before they start to generate hypotheses. For example, they might study the inner workings of some similar apparatus so they could draw analogies. Indeed, a whole series of black-box apparatus might be constructed with analogous mechanisms so that knowledge gained in the investigation of one could be utilized in the generation of hypotheses about others. This use of the apparatus would be much more in keeping with actual scientific practice.

Secondly, in the actual use of the apparatus students are given no time to mull over their ideas; they are given no rest, no incubation period. Indeed, in one case they are asked to come up with answers five or ten minutes after the problem is posed. An interesting idea would be to let students prepare by studying similar apparatus for a few days, and then only after several days or weeks have the students think up hypotheses. In the light of Wallas' partial method one might expect students to do better when incubation is allowed.

Of course, the pedagogical ideas presented here should not be taken on faith. Some interesting research in the teaching of science should be undertaken in connection with them. Thus research might investigate differences in the ability to generate hypotheses about the hidden mechanisms of the apparatus between students who were allowed to incubate and students who were not. Indeed, the approach to teaching hypothesis generation by means of black-box apparatus suggests even more important educational research. For example, educational research might investigate whether getting students to generate hypotheses by means of the black-box type of apparatus (either in the way suggested above or in the way used by the inventors) has any carry-over into more realistic scientific contexts. Do students who are given experience in producing true or at least plausible hypotheses about the inner mechanism of black boxes learn more easily to produce true or at least plausible hypotheses about the inner workings of the human body or the atom than those who do not receive black-box training? The answer to this question is by no means obvious.

THE CONTEXT OF TESTING

TWO APPROACHES TO THEORY TESTING

However a hypothesis is generated, the question still remains, is the hypothesis true? As we have seen, the truth of a hypothesis is not determined by the way it is generated. Thus, even if we know that the way a given hypothesis was generated is more successful than other ways, it will still be necessary to test the hypothesis. However, since we usually do *not* have knowledge of this sort, the necessity of testing the hypothesis is even more apparent.

When one tries to specify what the logic of testing is, however — what testing a hypothesis amounts to — a deep philosophical controversy arises. Two distinct views on the nature of hypothesis testing can be found in philosophical literature. According to the better-known view, hypothesis testing eliminates false hypotheses and establishes other hypotheses as probable relative to the available evidence. The more probable or well confirmed a hypothesis is, the more justified one is in supposing the hypothesis is true; and, other things being equal, the more justified one is in basing one's action on this hypothesis. The goal of science, in this view, is to arrive at a system of true hypotheses that enables one to explain, predict, and understand the world. The justification one has for believing that one has achieved such a system in part or in whole is the confirmation the hypotheses have in terms of the available evidence. Thus the context of testing involves gathering evidence that refutes or confirms hypotheses. The task of philosophy of science, in this view, is to clarify the sorts of inferences that are involved in this testing and the sort of evidence that provides support or refutation for hypotheses. This approach to hypothesis testing is often called the *confirmation approach*.[15]

There is another view of the matter, however. Some philosophers of science — most notably Karl Popper and his followers[16] — have argued that the idea that evidence supports, confirms, or makes probable a hypothesis is a mistaken one. The only function evidence has in hypothesis testing, they say, is to falsify or refute hypotheses. Since hypotheses are not supported, confirmed, or made probable by evidence, one cannot be justified in one's belief that a hypothesis is true on the basis of the evidence; moreover, one's actions cannot be justified on evidential grounds. The task of science, in this approach, is to set forth those hypotheses that are most easily falsifiable and to refute them. This view also holds that the goal of science is to achieve a system of true hypotheses that enables one to explain and predict and understand the world. But, unlike the confirmation approach, the present approach hopes to achieve this by the continued refutation of highly refutable hypotheses. The testing of a hypothesis therefore involves the gathering of evidence, but only in an attempt to refute the hypothesis, and the philosophy of science consists of clarifying the logic of refutation and refutability. This approach to hypothesis testing we will refer to as the *refutation approach*.

In considering these two approaches to hypothesis testing, let us consider in some detail the arguments used by Redi.

REDI'S TEST OF THE SPONTANEOUS GENERATION HYPOTHESIS

Redi argued that, if his hypothesis were true, then if fresh meat were closed in a glass flask, no maggots would appear. In addition, if spontaneous generation occurred, then the maggots would still appear in the closed flask. Moreover, if his hypothesis were true, then if fresh meat were placed in a glass flask which was not closed, maggots would appear.

Redi performed some experiments. He sealed some fresh meat in a glass flask and he put some fresh meat in a glass flask but did not protect this meat from the flies. As he expected, after several days no maggots appeared on the sealed meat, but maggots did appear on the unprotected meat.

Now one might argue that the spontaneous generation theory was still not disproven. One might maintain that spontaneous generation occurs only when there is free circulation of air, that Redi's sealing of the glass flask prevented air from circulating and thus prevented spontaneous generation from occurring. Redi performed another experiment, however, that would seem to silence this objection. He placed some meat in a flask and covered the flask with a fine veil. (This veil was supposed to protect the meat from the flies and yet allow air to circulate.) In another flask he put meat but did not protect it with a fine veil. The results of the experiment seemed to refute the modified version of the spontaneous generation hypothesis, since no maggots appeared in the flask protected by the fine veil and worms did appear in the flask unprotected by the fine veil.

On the basis of these experiments, Redi concluded that the worms were not spontaneously generated, but were the result of flies' laying eggs on the meat.

The arguments involved in Redi's investigation are typical in their general pattern of many arguments in science. Consider the following as the hypotheses tested by Redi:

> Maggots are spontaneously generated in meat.
> Maggots in meat are caused by flies' laying eggs on the meat.
> Maggots are spontaneously generated on the meat only when air circulates freely.

Let the following be test implications of these hypotheses, that is, sentences that seem to follow logically from these hypotheses. Thus

> If meat is placed in a sealed glass flask, then maggots will appear.

seems to follow from

> Maggots are spontaneously generated in meat.

And

> If meat is placed in a glass flask that is not protected against flies, then maggots will appear.

seems to follow from

> Maggots are caused by flies' laying eggs on the meat.

And

> If meat is placed in a glass covered with a fine veil, then maggots will appear.

seems to follow from

> Maggots are spontaneously generated on the meat only when air circulates freely.

Then Redi's arguments can be constructed as follows:

(1) If maggots are spontaneously generated in meat, then if meat is placed in a sealed flask, maggots will appear. But on the evidence, if meat is placed in a sealed flask, then maggots do *not* appear.

∴ Maggots are not spontaneously generated in meat.

(2) If maggots are caused by flies' laying eggs on the meat, then if meat is placed in a sealed flask, maggots will not appear.
As the evidence shows, if meat is placed in a sealed flask, then maggots do not appear.

∴ Maggots are caused by flies' laying eggs on the meat.

(3) If maggots are caused by flies' laying eggs on the meat, then if meat is placed in a glass flask that is not protected against flies, maggots will appear.
As the evidence shows, if meat is placed in a glass flask

that is not protected against flies, then maggots will appear.

∴ Maggots are caused by flies' laying eggs on the meat.

(4) If maggots are spontaneously generated on the meat only when air circulates freely, then if meat is placed in a glass flask covered with a fine veil, maggots will appear. But as other evidence shows, if meat is placed in a glass flask covered with a fine veil, then maggots will *not* appear.

∴ Maggots are not spontaneously generated on the meat only when air circulates freely.

The following points should be noticed: (i) Arguments (1) and (4) have the same form:

(A) If _____, then --------------------
     But not ----------------------------

     ∴ Not _____

Elementary logic teaches that all arguments with this form are logically valid. If the premises of an argument with this form are true, then the conclusion must be true. (ii) Arguments (2) and (3) have the following form:

(B) If _____, then --------------------

     ∴ _____

However, it is a well-known point of logic that such arguments are not valid. They commit "the fallacy of affirming the consequence." The premises of an argument with this form could be true while the conclusion could be false.

Philosophers advocating the confirmation approach to theory testing have argued that, although arguments like (2) and (3) are not deductively valid because they have form (B), they are still not without value; for example, the premises of the arguments in (2) and (3) support or confirm to a certain extent the hypothesis that maggots are caused by flies' laying eggs on the meat. It is important, however, to realize that even if a million test implications were deduced from the hypothesis that the maggots were caused by flies' laying eggs on the meat and were all confirmed by the evidence, this would not necessarily mean that this hypothesis was true. This is easily seen when it is remembered that there could be other conflicting hypotheses entailing the same test implications. Thus, even if the test implications of Redi's hypothesis are confirmed, there is no guarantee that his hypothesis is true — some other hypothesis might be compatible with the same evidence.

Philosophers advocating the refutation approach disagree with those advocating the confirmation approach over the value of arguments like (2) and (3). They maintain that the only legitimate arguments in science are those that are strictly deductive — arguments like (1) and (4). These arguments, it should be noted, *refute* the hypothesis being tested. Whereas the confirmation theorist would argue that (1) and (4) indirectly *confirm* or support Redi's hypothesis since this hypothesis has withstood the tests that the spontaneous generation hypotheses have failed, philosophers advocating the refutation approach would deny this.

AUXILIARY HYPOTHESES

We shall return to this controversy shortly. In any case, it is important to see that our account has been oversimplified in an important respect, given either the confirmation or the refutation account of testing. We have supposed that hypotheses tested in Redi's experiment are tested in isolation from other hypotheses. But this is clearly mistaken. A host of background assumptions play an important role in hypothesis testing.

We have already mentioned that someone might argue that air circulation was necessary to spontaneous generation. We had to assume that this was not necessary in argument (1). In (2) we tacitly assumed that flies could not get into a sealed flask to lay eggs. In (3) we supposed that flies could not get through a fine veil. In (4) we supposed that nearness to a fine veil does not prevent spontaneous generation. And so on. Thus it is clear that we were not testing these hypotheses in isolation. We were testing these hypotheses against a background of other assumptions, assumptions which may sometimes be difficult to articulate. In fact, arguments (1) and (4) must be amended as follows:

> (1′) If maggots are spontaneously generated on meat and certain background assumptions are true, then if meat is placed in a sealed flask, maggots will appear.
> But on the evidence, if meat is placed in a sealed flask, then maggots do not appear.

∴ Either maggots are not spontaneously generated on meat, or one or more of the background assumptions is false.

(4') If maggots are spontaneously generated on meat only when air circulates freely and certain background assumptions are true, then if meat is placed in a glass covered with a fine veil, maggots will appear.
But if meat is placed in a glass covered with a fine veil, maggots will not appear.

∴ Either maggots are not spontaneously generated on meat only when air circulates freely, or one or more of the background assumptions is not true.

Similar changes would have to be made in arguments (2) and (3) used by the confirmation approach.

When arguments (1) and (4) are amended to (1') and (4') we see that Redi's experiments do not necessarily refute the spontaneous generation hypothesis. For it might be that one or more of the background assumptions was false. Thus in (4') the hypothesis might be true, but the background assumption that a fine veil near meat does not prevent spontaneous generation might be false. But then the fact that the test implication is false might be due to the fact that the background assumption is not true, rather than that the hypothesis is not true.

The account of hypothesis testing given so far is still incomplete in at least one respect. It is not always easy to deduce test implications from a hypothesis and the background assumptions usually associated with it. Hypotheses and background assumptions in science are often vaguely or ambiguously stated and it is sometimes difficult to tell whether a certain proposition follows from them; hence, it is difficult to tell whether the testing of the proposition is relevant to the testing of the hypothesis.

This problem—one that is too often neglected in standard accounts of hypothesis testing—is usually handled in practice in the following way. The vague or ambiguous terms are reinterpreted or reconstructed in a way that enables one to decide whether a proposition is a test implication of the theory or not.

Thus, a vague hypothesis and background assumptions sometimes entail a test implication only when some of the terms in which the hypothesis is stated are given a certain reconstruction. One might consider this reconstruction simply another background assumption, but to do so would be misleading. This reconstruction is not a factual assumption about the world; it cannot be refuted or confirmed. The reconstruction is more like a recommendation or proposal to use terms in a way that is fruitful for scientific purposes. (We will consider such reconstructions of scientific terms in detail in Chapter Three.) After a while such recommendations become common practice: the reconstruction is absorbed into the actual meaning of the scientific terms and no reconstruction is needed to draw test implications from a hypothesis in which the term figures.

So one might say, then, that in some scientific contexts—at least in the early stages of hypothesis testing—the inference patterns outlined above should be changed. The general form would be something like this:

> If a certain hypothesis and background assumptions are true, given a certain reconstruction, then a certain test implication would be true.
> But the evidence shows that this test implication is not true.
>
> ---
>
> ∴ Either the hypothesis is not true or one or more of the background assumptions are not true, given the reconstruction.

Thus, in the case of negative evidence, one could modify the hypothesis or modify one or more of the background assumptions or modify the particular reconstruction one has given the terms in the hypothesis and the background assumptions in order to avoid a conflict between the evidence and the hypotheses.

TESTING AND EXPERIMENTATION

One widely held view of science is that experimentation is a necessary condition for testing. This view is mistaken. Doing an experiment involves bringing something about deliberately in order to observe a change in something else. In our terms experimentation would be possible when the antecedent of the test implications of a hypothesis specified something that could be brought about. Redi's test of the spontaneous generation hypothesis was an experiment since the antecedents of the test implications all specified things that could be brought about.

However, not all testing of hypotheses is of this kind. Sometimes the antecedent of the test implications of a hypothesis does not specify something that can be brought about—at least in the present state of our technology. The test implications of some hypotheses of astronomy are of this kind. Consider, for example, Kepler's third law of planetary motion, which states that if T is the time for a complete orbital revolution of a planet about the sun and $\bar{R}$ is the mean radius of the orbit of that planet, then $T^2 = k\bar{R}^3$ where k is a constant quantity that has the same value for all planets. This law combined with certain background assumptions has test implications. But the antecedents of these test implications cannot in general be brought about by human design since they would refer to some specific period or radius of a planet — something that cannot be created or manipulated at the present stage of technology. Thus new direct empirical tests of Kepler's law are not achieved through experimentation but in other ways. The discovery of a new planet, Pluto, provided such a direct test since the new planet's period and mean radius could be checked against the law. Moreover, as we shall see later, this law is also subject to indirect tests. However, as our technology increases, direct experimental testing may become possible. Artificial satellites may be placed in various orbits around the sun and their periods observed.

It should be clear from this example not only that experimentation is not a necessary condition for testing hypotheses, but that experimentation is not found in all branches of the natural sciences. The widely held view that the natural sciences are exclusively experimental sciences is thus mistaken. Moreover, not all hypothesis testing in the social or behavioral sciences is nonexperimental. In certain branches of psychology, for instance, experimental testing of hypotheses is widespread.

There is another mistaken notion about testing and experimentation that should be mentioned. Some have supposed that in order to perform a test one must hold all factors constant except one. Consider, for example, the law of the lever:

$$\frac{F_2}{F_1} = \frac{l_1}{l_2}$$

To put this into words, the ratio of the load to the applied force is equal to the inverse ratio of the respective lever arm. A statement of this kind would yield an indefinite number of test implications, some of which have the following form: If l_2 is the length of the lever arm from the fulcrum to the load and l_1 is the length of the lever arm from the fulcrum to the applied force and F_2 is the weight of the load, then F_1 is the amount of the applied force necessary to lift the load. An experimental test then might consist of varying the value of l_2, holding l_1 and F_2 constant, and observing the change in F_1. However, many other factors change during this process and it is quite clear that all of them could not be held constant: the distance of the lever from Mars, the physiological condi-

tion of the experimenter, the movement of the air in the laboratory, and so on. If it were necessary to hold all factors constant
except one, experimentation would be impossible. But the law
and background assumptions imply that these other factors are
irrelevant and thus there is no reason to keep them constant.
Moreover, as we shall see shortly, a case can be made for not holding these other factors constant in testing the law.

CRITERIA FOR CONFIRMATION

For advocates of the confirmation approach to theory testing
the question remains: What increases the evidential support or
confirmation of a hypothesis? The answer usually is given in
terms of four factors: (1) amount of evidence, (2) variety of evidence, (3) precision of evidence, and (4) indirect theoretical support.[17] Let us consider these factors in turn.

(1) If there is no negative evidence, the more confirmed consequences of a hypothesis there are the more support the hypothesis receives. Thus Redi's hypothesis received more support from
the results of his experiment with the sealed flask and the veil-
covered flask than from either alone. In the case of the lever law,
the greater the number of test implications that are deduced from
the law and confirmed the more support the law receives.

(2) However, the *amount* of evidence supporting a hypothesis
is often not considered as important as the *variety* of the evidence.
Testing the lever law, for instance, under a variety of conditions
makes it more likely that the law, if it is false, will be refuted by
the evidence. (Here we see the importance of not holding all factors constant even if one could, since holding all factors constant
would be tantamount to not having great variety in the evidence.)
Thus, if the law withstands the test, it receives strong support.
Indeed, some inductive logicians have argued that an increase in
the amount of evidence increases the confirmation of a hypothesis
only because of an increase in the variety of the evidence that ac-
companies the increase in amount.

In any case, it is true that the most important theories in science have withstood tests in a variety of different circumstances.
Thus Newton's theory entails a number of laws in diverse areas:
laws for planetary motion, laws for the simple pendulum, laws for
free fall, and so on. The confirmation of these laws, it is argued,
provides strong support for the theory.

However, variety per se does not always increase the confirmation of a hypothesis. This is true when the variety is likely, in
the light of background information, to be irrelevant to the hypothesis. The eye color of the experimenter, the number of hairs
on the experimenter's head, the birth date of the experimenter's
wife would be judged irrelevant factors in the experiment per-

formed by Redi both in the light of our background information and undoubtedly in the light of his background information. Thus experiments in which irrelevant factors were varied would not be thought to increase the support of the hypothesis.

(3) Another factor that is supposed to increase the support of the hypothesis is the *precision* of the evidence. Consider Kepler's third law. Suppose we deduce from the law, given $\bar{R}$ and k, the period T of Jupiter. The test is positive. Suppose the observed values $\bar{R}$ and T were based on some observation method M_1. Suppose, now, that a more exact measuring procedure, M_2, is used and our results are still positive. It is thought that the evidence confirms the law more in the latter case than in the former. The reason given is similar to the reason given for variety in the evidence increasing the degree of support. More precise measurement techniques make it more likely that, if the hypothesis is false, it will be refuted by the evidence. If the hypothesis withstands the more stringent test, it is thereby more strongly supported.

(4) In the confirmation approach to hypothesis testing, a hypothesis can also receive support indirectly. One important way of providing indirect support is by showing that the hypothesis (or a close approximation) is a deductive consequence of a more general hypothesis that is independently well-supported. Thus Kepler's law received indirect support from Newton's theory which was independently well-supported; Newton's theory has deductive consequences of which Kepler's laws are close approximations.

The converse can also occur. A hypothesis can become improbable because it conflicts with well-established theory. Consider a hypothesis that was fundamental to the Bates system of sight therapy. Dr. William Bates and his followers argued that accommodation, the focusing process which takes place within each eye when attention is shifted to objects at various distances, is brought about by two muscles on the outside of the eye. These muscles alter the entire length of the eyeball. On the basis of this hypothesis, Bates argued that glasses were unnecessary and even harmful in the correction of refraction problems; he substituted a system of eye exercises and other methods. Bates and his followers claimed many cures using this system, but most oculists rejected his views. It is a well-established hypothesis of eye anatomy that accommodation is the result of changes in the lens caused by the movement of the ciliary muscles. Eye exercises and other Batesian methods do not affect this muscle. Bates' "cures" must, therefore, be explained in other ways or be dismissed as specious.[18]

The indirect way of confirming a hypothesis indicates that there is a built-in conservatism in science, at least given the confirmation approach. The old established theory has a strong *prima facie* claim to truth and the burden of disproof is on the challenger to the established order. Indeed, sometimes the lack of adequate explanatory hypotheses for evidence that seems to challenge the established system of hypotheses keeps the evidence from being scientifically acceptable. This seems to be the case with evidence resulting from ESP experiments. Although some of this evidence seems uncontestable, it is in conflict with well-established theories. Until an adequate explanation of these results is forthcoming, this evidence remains suspended in a scientific limbo, waiting to be redeemed, and the system of established hypotheses is in a temporary state of grace.

TYPES OF NONDEMONSTRATIVE INFERENCE

We have considered so far two inferences used in scientific investigations. These illustrate two types of inference: demonstrative and nondemonstrative. The inference used in the refutation approach, namely,

If _____, then --------------------
Not --------------------

∴ Not _____

is a particular form of valid demonstrative inference. In valid demonstrative inferences the premises necessitate the conclusion — if the premises are true, the conclusion must be true. Inferences of this sort are used in formal logic and mathematics. Their major role in science is in articulating the logical consequences of a hypothesis, thus permitting the hypothesis to be tested.

The inference mentioned in connection with the confirmation approach, namely,

If _____, then --------------------

∴ _____

is a particular form of nondemonstrative inference. Let us call this a *hypothetical inference*. In hypothetical inferences, the premises can be true and the conclusion false. Nevertheless, as we have seen, advocates of the confirmation approach to hypothesis testing believe that in these inferences the premises provide support for the conclusion. They also argue that some other nondemonstrative inferences do this as well.[19]

One of the most common nondemonstrative inferences is called *inductive generalization*. One draws from the fact that all observed members of the sample have a particular property the conclusion that all members of the population have such a property. The degree of confirmation afforded the conclusion is relative to the amount and variety of the evidence constituted by the sample. Thus, if copper observed in wide and various circumstances conducts electricity in every case, this lends support to the hypothesis that all pieces of copper conduct electricity. However, it should be clear that this inference is nondemonstrative for there is nothing contradictory in the supposition that all observed pieces of copper conduct electricity and that some unobserved pieces do not.

A similar inference—sometimes called a *statistical generalization*—is an inference from the fact that a proportion of a sample has a property to the conclusion that the same proportion of the whole population does. Thus, for example, the fact that two thirds of the people in the United States with an income of four to seven thousand dollars in wide and various circumstances were found to visit a doctor at least once a year lends support to the conclusion that two thirds of all the people in the United States in this income bracket visit a doctor at least once a year.

Some nondemonstrative inferences proceed from the examined members of a class to some unexamined members. Let us call these *predictive inferences*. Thus the fact that all sodium salt observed in a wide variety of circumstances burns yellow lends support to the hypothesis that the next piece of sodium salt examined will also burn yellow.

Other nondemonstrative inferences, which we shall call *direct inferences*, proceed from premises about the makeup of the whole population to some sample drawn at random from the population. Thus the fact that 15 percent of the Hindus in India are of the Harijan caste confirms the conclusion that a sample of Indian Hindus generated by some random selection method would have a similar proportion of Harijan. Similarly this fact would lend support to the conclusion that some particular Hindu picked at random would not be of this caste.

THE JUSTIFICATION OF NONDEMONSTRATIVE INFERENCE

So far we have considered several nondemonstrative inferences in science, inferences in which the premises are supposed to support or confirm the conclusion of an argument, although they do not logically entail it. The question arises: Why should we rely on these inferences? The advocates of the confirmation approach to theory testing must make some defense, for they have surely supposed we should rely on these inferences.

This problem—the justification of nondemonstrative inferences—was first explicitly posed by David Hume and it can be put rather simply: Nondemonstrative inference is by definition inference in which 'he premises do not necessitate the conclusion. Consider, for example, the premise "The sun has always come up in the past" and the conclusion "The sun will come up tomorrow." If there is no logical connection between the premise and the conclusion, how are we to justify the conclusion in terms of the premise? What reason do we have to believe, for example, that the sun will come up tomorrow? The argument that the premises make the conclusion probable, Hume maintained, presupposes that the future will resemble the past. But this presupposition is an assumption about matters of fact, hence it is not a necessary truth. How then is it to be established? Surely not on the basis of the nondemonstrative inferences considered above when they are used to infer future events; these nondemonstrative inferences also presuppose that the future will resemble the past.[20]

Most philosophers have found Hume's answer to this problem—often called the problem of induction—hard to accept. Hume maintained that we cannot in any way show that the future will resemble the past. Our belief in the reliability of the nondemonstrative inferences outlined above, he said, is based upon custom or habit and has no rational foundation.

Let us consider briefly some of the problems involved in showing that the sorts of nondemonstrative inferences considered above are justified, that these inferences can be relied upon.[21]

Let N be the nondemonstrative inferences considered above. Let us suppose that in the past the use of N has usually been successful. "Usually successful" might mean different things depending on the type of inference. In the case of predictive inference something like this would be meant: Predictive inferences which have been based on large and varied amounts of evidence have usually led to true conclusions. In the case of hypothetical inferences "usually successful" might mean: The conclusions of hypothetical inferences that have been based on large, various, and precise evidence have usually not been refuted by subsequent tests. Now one might be inclined to argue in this way:

> (1) In the past the use of N has been observed to be usually successful on numerous occasions in a wide variety of circumstances.

(1) lends support to

> (2) The next uses made of N, e.g., in the next twenty years, will usually be successful.

However, it should be clear that this argument is question begging. In arguing that (1) supports (2), a particular nondemonstrative inference is assumed, namely a predictive inference. But the justification of such inferences is precisely what is in question. Thus this attempt to justify nondemonstrative inference fails.

Some recent philosophers have argued that the traditional attempts to justify the types of nondemonstrative inference used in science should be given up since the justification of nondemonstrative inference—the problem of induction—is a pseudoproblem. The use of certain nondemonstrative inferences is simply part of what is meant by being rational, they argue. These inferences and the criteria governing their use provide our standards for evaluating particular nondemonstrative inferences. They grant that it does make sense to ask for a justification for *particular* nondemonstrative inferences. We judge particular inferences in terms of the standards of nondemonstrative inferences used in science. Consider the particular nondemonstrative inference from

(1) All observed copper conducts electricity

to

(2) All copper conducts electricity.

Does (1) support (2)? Yes, since the inference is an inductive generalization that is based on a large amount of evidence that has great variety. These are the relevant standards in this case. And this is all that it means for (1) to support (2). However, the attempt to justify these standards themselves makes as much sense as trying to give a legal justification of the United States Constitution. The Constitution is the standard for all law in the United States and it thus makes no sense to ask for a justification of it.

However, while it does not make sense to ask for a *legal* justification of the United States Constitution, we can ask for a *moral* justification; we can ask why we *ought* to follow the Constitution. Similarly we can ask why we *should* use the standards we do use in judging particular nondemonstrative inferences. Both questions seem to make sense. Thus this attempt to show that the problem of the justification of nondemonstrative inferences is a pseudoproblem fails.

Although there have been many attempts to justify the sorts of nondemonstrative inferences used in science, none of them is completely successful. However, an adequate justification may yet be found, since no one has ever shown to the satisfaction of all competent students of the subject that such a justification is logically impossible. Much depends on what *sort* of justification it would be appropriate to ask for. And this is by no means clear.

In any case, this failure of the confirmation approach to hypothesis testing to find a justification for nondemonstrative inferences does indirectly seem to support the refutation approach. This approach, it will be recalled, maintains that hypotheses are not supported or confirmed by the evidence. It does not deny that commonly accepted standards of nondemonstrative inference are applied to particular inferences. Rather it denies that there is any good reason to rely on these standards.

The refutation approach to science, however, has problems of its own. Popper and his pupils have made refutation and refutability the central concepts in science. As a result they have maintained the following theses:

(1) Those hypotheses that are most refutable but have not yet been refuted are always to be preferred to those that are less refutable.
(2) Those hypotheses that are most refutable are the ones that are most simple.
(3) Refutability is a sufficient and necessary condition for separating science from nonscience.[22]

Let us consider these theses.

It does not always seem to be true in science that the most refutable but as yet unrefuted hypothesis is the one that is the most desirable. Other things being equal, the competent scientist seems to prefer the simplest hypothesis compatible with the evidence. But the simplest hypothesis is not necessarily the most refutable.

Suppose we examine a number of pieces of copper in our laboratory over the Fourth of July weekend. All the pieces examined conduct electricity. Let us call this evidence e. We conjecture that:

h_1 All copper conducts electricity.

However, other hypotheses are compatible with our evidence e, for example:

h_2 All copper and large pieces of iron from Ohio conduct electricity.

Although h_2 is more easily refutable than h_1, surely h_1 is the preferable hypothesis relative to evidence e. Moreover, h_2 is less simple than h_1. Indeed, it is plausible to suppose that h_1 is preferable to h_2 because h_1 is simpler than h_2.

This, of course, does not mean that the most preferable hypothesis is the least refutable. Consider another hypothesis:

h_3 All pieces of copper in the United States over the July Fourth weekend conduct electricity.

Although h_3 is less refutable than h_1, it would surely not be preferable to h_1. Thus it seems to be a mistake to identify simplicity with either high refutability or low refutability. Popper then seems to be mistaken about (2). And insofar as the most preferable hypotheses are the simplest hypotheses — other things being equal — he is mistaken about (1).[23]

Moreover, Popper seems to be mistaken about interesting special cases of (2).[24] Consider:

(2′) The most refutable hypothesis about a planetary path is always the simplest hypothesis about the path.

It is generally agreed that a hypothesis that specifies that a planet travels in a straight line is more simple than a hypothesis that specifies that a planet travels in a circle. A hypothesis that specifies that a planet travels in a circle is more simple than a hypothesis that specifies that a planet travels in an ellipse. And so on for higher degree curves.

Popper argues that, if it takes n number of points to specify the path of a planet, then it takes $n + 1$ number of observations to refute the corresponding hypothesis. Thus it would take two points to specify the path of a planet that travels in a straight line and three observations to refute this hypothesis. Three points would be necessary to specify the path of a planet that travels in a circle and four observations to refute this hypothesis, and so on. Thus, the more complex the curve the harder it is to refute.

This contention seems to be mistaken. It would take only three observations to refute the hypothesis that a planet travels in a straight line, as well as to refute the hypothesis that it travels in a circle or an ellipse. In the case of the circle hypothesis and the ellipse hypothesis, three observations that could be joined by a straight line would be all that would be necessary for refutation; in the case of the straight-line hypothesis, three observations that did not form a straight line would suffice. In short, the degree of refutability is the same although the degree of simplicity differs. Popper seems mistaken in his general thesis (2), as well as in (2′), a special case of (2).

Although Popper's analysis and justification of simplicity in terms of refutability seems to be mistaken, it should not be supposed that a completely adequate analysis and justification of simplicity is available. Despite work by competent scholars, the concept of simplicity remains in large part inadequately analyzed and the use of simple hypotheses in science remains unjustified.

Now, although Popper may be wrong about (1) and (2), he may still be correct about (3). He has maintained that one of the fundamental problems in the philosophy of science is to distinguish science from nonscience. He has stressed that he is not concerned with the problem of the logical positivist: the separation of meaningful discourse from nonmeaningful discourse. Thus, when Popper speaks of the demarcation between science and nonscience, he does not mean what the logical positivists mean by nonscience, nonsense; he means meaningful nonscience.

Popper proposes as a way of differentiating science from meaningful nonscience the refutability of the former and nonrefutability of the latter. Popper seems to take refutability as both a necessary and a sufficient condition for distinguishing (a) scientific propositions and (b) scientific systems from nonscientific systems. Thus, according to Popper:

(a) p is a scientific proposition if and only if p is refutable.

Such a view has the paradoxical consequence that certain existential propositions, i.e., propositions asserting the existence of something, would not be scientific. For example:

p_1 There is at least one object that is radioactive.

p_1 is a hypothesis which we would normally call scientific, but which cannot be refuted since no amount of possible evidence could be in conflict with it. In this respect it is unlike negative existential statements—ones that assert that something does not exist—and existential statements that specify some particular time and place. Consider:

p_2 There is not even one object that is radioactive.
p_3 There is at least one object that is radioactive in this box now.

For p_2 and p_3 evidence could be specified which would refute them. Now Popper does not seem to mind saying that certain propositions like p_1 are not scientific. He would maintain, however, that this only shows that there are certain "metaphysical," i.e., nonscientific, statements that are part of a system of scientific hypotheses. He would hold that the total system is refutable even if some parts of it, taken by themselves, are not; thus the major concern of Popper is to distinguish a scientific system of hypotheses from a nonscientific system. He maintains:

(b) A system S of hypotheses is a scientific system if and only if system S is refutable.

However, it is dubious that such a criterion provides a sufficient condition for distinguishing a scientific system of hypotheses from a nonscientific system. For if S is a refutable system of hypotheses, then so is S' where $S' = S$ and M (where M is any proposition at all). But suppose M is "God is good." To say that S', which contains the statement "God is good," is a scientific system of hypotheses is surely paradoxical. Thus (3) seems dubious.[25]

There is one final critical question that should be mentioned: Why should one use Popper's method? Why should one put forth refutable hypotheses and subject them to empirical tests? There are two possible replies that Popperians could make:

(4) Attempting to refute highly refutable theories will lead to truth.
(5) Attempting to refute highly refutable theories will eliminate error.

Indeed, one might argue that (5) is a good reason for accepting (4).

These replies, however, seem to be unjustified. There is no *a priori* or logical reason to suppose that (4) is true even if one grants (5), since (5) does not entail (4). Suppose that one had refuted a number of highly refutable theories. This would not necessarily mean that any true hypotheses of the relevant kind were known. If one eliminates hypothesis H as false, it does not necessarily mean that one knows any true hypotheses (except $-H$, the formal negation of H). Nor, on Popperian ground, is there any *a posteriori* or empirical reason to believe that (4) is true on the basis of (5). An empirical reason in this case would presumably be some sort of inductive argument. One might argue, for example, from the premise that elimination of false hypotheses has led to true hypotheses in the past to the conclusion that probably this will be so in the future. However, this conclusion cannot be accepted, according to the Popperian view, since no inductive conclusion is justified.

(5) does not fare much better. It may be correct that attempting to refute highly refutable theories has eliminated error in the past. But if Hume is correct, this gives no reason to suppose this will continue to be so. No refutations of new refutable hypotheses may be forthcoming—no matter how hard scientists try to refute them. However, this will give us no ground, according to Popper's view, to suppose that the hypotheses are not false. Perhaps the most that can be said is:

(5′) Attempting to refute highly refutable theories can eliminate error in the future.

And this seems to be true by definition of the term "refutable." It could also be said that:

(4′) Attempting to refute highly refutable theories could lead to truth.

But there is a crucial difference between (4′) and (5′). In Popper's view, we would have no reason to think that we had arrived at true hypotheses in the future even if we had arrived at them, since failure to refute a hypothesis gives us no reason in Popper's system to suppose the hypothesis is true. But we would have good reason to suppose we had eliminated error if we had eliminated it, since a false consequence derived from a system of hypotheses is deductive proof that one or more of the hypotheses in the system is false. Thus (5′) and not (4′) has methodological interest and provides a rather weak justification for the use of the Popperian method: use of the method guarantees that error can be eliminated in science and that scientists will have deductive grounds for supposing that error has been discovered when and if it has been.

This extended critical discussion of the refutation approach can perhaps be summarized in this way. Popper's simple identification – high refutability = simplicity = desirability – is mistaken. Some hypotheses are preferable to and simpler than other hypotheses which are more refutable. Popper's attempt to distinguish science from nonscience in terms of refutability seems to fail. Certain obviously scientific statements are not refutable and hence not scientific in his account; certain clearly nonscientific systems are made refutable, hence scientific in his account. Finally, the reason for adopting the Popperian viewpoint is a rather weak one. It cannot be the probability of finding truth or eliminating error if one uses this approach. The reason seems only to be that the knowledge that error has been eliminated is possible.

However, whatever the philosophical problems connected with the refutation approach, this approach does suggest a new way of looking at science education. We shall see how in a moment.

SCIENCE EDUCATION AND TESTING

SCIENCE EDUCATION AND THE CONFIRMATION APPROACH

Science educators generally assume without question that hypotheses in science are confirmed, that hypotheses can become more probable relative to certain evidence. As a result, the refutation approach has never, to my knowledge, been incorporated into science education or adopted as an approach to science education. This is unfortunate.

Equally unfortunate is the fact that science educators who tacitly adopt the confirmation approach to theory testing present to their students an extremely simple view of what the confirmation of a hypothesis involves. Indeed, the view of confirmation presented in a typical science education book is not in keeping with the general level of scientific sophistication presented in the same book. Put in a different way, rather complex and difficult scientific theories are presented with some sophistication along with a simple-minded view *about* scientific theories and evidence. (Indeed, sometimes these views about theory testing are in conflict with the illustrations used in the book.) Surely if science students can be expected to master the theory of relativity, the theory of evolution, and modern theories of heredity, they can be expected to grasp more sophisticated views about theory confirmation.

For example, consider Redi's experiment. This experiment has also been discussed by science educators, but there are rather basic deficiencies in their accounts of it. Consider, for example, the discussion of Redi in the Biological Science Curriculum Study's (B.S.C.S.) *Biological Science: An Inquiry in Life.* It is argued:

> What he [Redi] had shown was that under the conditions of his experiment maggots did not arise spontaneously in decaying meat.[26]

However, as we have seen, this is not so. Redi's evidence was disconfirming only relative to certain auxiliary hypotheses about how spontaneous generation works. Indeed, there is no explicit discussion of the role of auxiliary hypotheses in Redi's experiment at all in *Biological Science.* Yet without such a discussion the tentativeness of the results—stated in the text—remains completely obscure. The whole discussion of Redi's experiment in this book would have been enormously clarified by the explicit use of different types of logical arguments and a little formalism: the very sort that was utilized above in our philosophical analysis of theory testing.

Biological Science also notes that "Redi did feel justified" in supposing that life was formed by reproduction and not spontaneous generation. It remains unclear why Redi *should* have felt justified by his evidence. There is no explicit discussion of the types of factors that supported Redi's hypothesis, no explicit discussion of the criteria of confirmation. Moreover, when such factors are hinted at in the text they seem to be too simple. We learn that what was needed with respect to Redi's hypothesis "was *con-*

firmation. Others must repeat his experiment and do similar experiments. If the same results were obtained, then Redi's general hypothesis . . . would be more secure."[27] From this passage one might guess that the *amount* of evidence is important in confirmation and that the variety of evidence, more precise evidence, and rather different kinds of experiments is of no importance. But as we have seen, just the reverse is true. Indeed, the criteria of confirmation implicitly assumed in the text are belied in the illustration in the text. The experiments of Joblot, Needham, and Pasteur cited and discussed in the text to illustrate the further confirmation of Redi's hypothesis were not repeats of or even similar to his experiment; they were experiments that yielded a wide variety of evidence, and more precise evidence. This point is completely obscured by the view of confirmation held by the authors.

Another typical view about confirmation is presented in the Earth Science Curriculum Project's (E.S.C.P.) *Investigating the Earth.* The authors say:

> What is the difference between evidence and proof? . . .
> *Evidence* is an observation that tends to support a conclusion. A *conclusion* is an interpretation or judgment based on the evidence. When there is a little evidence for it, a conclusion is only probable. The conclusion becomes proved when there is sufficient evidence to support it.[28]

The difference between evidence and proof is left completely obscure by this passage. What does it mean to say that a conclusion becomes proved when there is sufficient evidence to support it? What does "sufficient" mean here? Does it mean that certain evidence will give us certainty? But there is no certainty in science. Does it mean merely that when there is a proof the conclusion is very probable relative to the evidence? If so, talk about proof seems to obscure this.

More importantly, there is no discussion of what factors constitute proof, e.g., when a hypothesis is very well supported by the evidence rather than merely supported. This is especially important for students using this book, since the authors go on immediately to ask whether certain data are evidence or proof that the earth is round. It is unclear how students could possibly answer such a question unless the authors clarify what distinction they have in mind and what criteria they are using in making the distinction.

SCIENCE EDUCATION AND THE REFUTATION APPROACH

An approach to science education that emphasized refutation would have a rather different slant than any science program or science curriculum today. First, in studying cases from the history of science, the refutation of past theories would be stressed. The history of science would not be presented as it often is, as a series of theories that become more and more confirmed. For example, Redi's hypothesis is usually presented as becoming more completely supported by later investigators until it is known to be true. Instead the stress would be on past theories that were overthrown, on the modifications in Redi's hypothesis that had to be made in the light of negative evidence, on the tentative character of the hypotheses, and on the evidence and experiments that might yet refute it. In short, the history of science would be viewed as a history of conjecture and refutation, as a history of how scientists learned from their past mistakes and how they might continue to learn.

Secondly, students would learn to refute theories. The questions constantly asked in class would be "What evidence would count against this view?" "How could this evidence be obtained?" "What procedures would most severely test this view?" The laboratory would not have as its function the illustration of certain physical principles; it would be a place where students learned to subject theories to test and to possible refutation.

Thirdly, science textbooks would be written with the refutation approach in mind. The stress would be on what evidence would overthrow theories, what would be the most severe tests of the theories studied. The typical sorts of questions found at the end of chapters might be different or at least be given a different interpretation. Consider, for example, a question from *Biological Science:*

(1) What evidence and data support a history of evolution for man, as for other organisms?[29]

Rewritten with a Popperian slant the question might become:

(1') What evidence or data would refute a history of evolution for man, as for other organisms? What evidence has refuted antievolutionary histories of man?

Some questions could be given a Popperian interpretation and answered without any change of wording. The answers wanted by the text, however, would not be accepted on Popperian grounds. Consider for example:

(9) You have read that a great deal of controversy existed as biologists conducted experiments relating to the origin of life. Controversy of this sort exists today in all branches of biology and is considered to be a vital ele-

> ment in the development of biology knowledge. Why is
> the controversy so valuable?[30]

Clearly the answer the text wants is in terms of medical practice, for it maintains that "preventive medicine becomes possible
with biogenesis." However, Popperians would answer the question
somewhat differently. They would hold that controversy is vital
not for some practical reason, but because it will stimulate severe
testing and the refutation of false hypotheses. In this way, according to Popperians, science progresses.

THE USE OF PSEUDOSCIENCE IN SCIENCE EDUCATION

Whether the confirmation or refutation approach is adopted
by science educators, one thing seems clear. Teaching students
the importance of testing scientific theories is crucial. One pedagogical way of stressing the importance of testing in science is to
contrast science with other areas which do not stress testing. A
particularly illuminating contrast may be made between science
and pseudoscience. A pseudoscience is a systematic body of propositions, practices, and attitudes that gives the appearance of being a science but is not a science. This means that a pseudoscience
has certain properties that are characteristic of many sciences and
give it the appearance of a science, but other properties, often not
easily discernible, that are not characteristic of a science. The discovery of these latter properties reveals the unscientific nature of
the body of propositions, the practices and attitudes that constitute a pseudoscience.

Let us call the properties which give pseudoscience the appearance of a science the *surface properties*, and the properties
which are hidden but which reveal the unscientific nature of the
body of propositions, practices, and attitudes the *depth properties*.
Among the surface properties of a pseudoscience are usually
these: The propositions of the pseudoscience will be couched in
technical language that will be used to express far-ranging and
impressive-sounding theories. The pseudoscientist will claim that
these theories are well supported by the evidence and will use
ingeniously complex and impressive arguments to meet criticism
of his theories. Pseudoscience might include special training for
the practitioners, special organizations and journals patterned
after scientific professional organizations and journals, and the
use of an authoritative text. Among the depth properties of the
pseudoscience are these: The propositions that make up the pseudoscience are either untested or untestable, or perhaps already
refuted. The pseudoscientist attempts to prevent his propositions
from being exposed to critical test and evaluation, and explains
away any possible negative evidence. The pseudoscientist or

group of pseudoscientists isolate themselves from the mainstream of scientific practice and from critical interaction with the scientific community. The attitude of the pseudoscientist is dogmatic and slightly paranoid; he is intolerant of all theories except his own.[31]

One might say that the basic methodological difference between pseudoscience and science lies in what Schwab has called the *syntactical structure*, that is, in the differences in the canons and standards of proof and evidence used in pseudoscience and science.[32] Science critically tests its theories and hypotheses and modifies them in the light of the evidence; pseudoscience does not. Moreover, although the surface properties of pseudoscience and science are the same, they have quite different functions that reflect the basic differences in the depth properties. Science's technical vocabulary, journals, and professional organizations have as one of their primary functions the furthering of the critical spirit of science. For example, technical vocabulary is used to state hypotheses more precisely so that more severe testing is facilitated; professional journals not only publish new scientific discoveries but enable critical responses to be voiced. Even authoritative texts are subject to critical review and are replaced as science progresses. In pseudoscience, technical vocabulary has as its function obscuring what is being said, thus preventing testing and criticism; journals (if they exist) function to reinforce the dogmas of the pseudoscientists who contribute to them. The authoritative text in pseudoscience is authoritative much in the way the Bible is authoritative to fundamentalists — it is not a framework to critically evaluate, test, and perhaps overthrow.

Despite the fact that the study of pseudoscience would illuminate the study of scientific theory testing, science educators have in general neglected this area. Indeed, science educators seem almost to have a strong aversion to mentioning pseudoscience in textbooks and curriculum material; detailed discussions of pseudoscience in science education material are virtually unknown.

Perhaps science educators avoid pseudoscience because they believe that if students studied pseudoscience they would become pseudoscientific. This belief is similar to the views of right-wing activists who are opposed to studying communism in school because they believe that such study would make students communists. However, as is well recognized, the objective study of communism in school is perhaps the best guard against communism. This is undoubtedly also true in the study of pseudoscience. Students who study pseudoscience in the proper way will probably be less likely to be taken in by the claims of pseudoscience than students who do not. But, it may be asked, what is the proper way?

In the case of the study of communism, there is a distinction between teaching students *to be* communists and teaching students *about* communism. In a similar way, there is a distinction between teaching students *to be* pseudoscientists and teaching *about* pseudoscience. Furthermore, there is a distinction between teaching about pseudoscience from a historical, sociological, or psychological point of view and teaching about pseudoscience from a methodological and critical point of view. Surely it is the latter that is of most interest for science education.

There are many ways in which the study of pseudoscience could be incorporated into science education:

(1) Students might critically study historical cases of pseudoscience along with cases of genuine science. Thus the Conant historical case study method could be easily adapted to the study of pseudoscience.[33] For example, some papers of Lysenko could be read in conjunction with Mendel's original papers.

(2) Contemporary research papers and contemporary pseudoscientific "research" papers—preferably on the same topic—could be read.[34] Differences in the use of evidence and hypotheses and in the methodological attitudes manifested in the papers could be brought to light. For example, papers on the "cure" of some particular disease from the journal of the National Chiropractic Association could be read in conjunction with research papers from medical journals on the cure of the disease.

(3) Laboratory work could be directed in part to exposing the scientific pretensions of pseudoscience. For example, some of the pseudoscientific theses of Charles Wentworth Littlefield and Morely Martin could be tested in the lab.[35]

(4) Students might be encouraged to bring in for class discussion examples of pseudoscientific thinking, beliefs, and theories that they find in newspapers, magazines, and other media. For instance, advertisements of food faddists could be critically discussed.

(5) Textbooks might be written with at least a chapter devoted to a critical consideration of some pseudoscientific theory. For example, a biology text might consider critically the pseudoscientific theory of accommodation presented by Dr. William Bates and contrast it with the theory of accommodation accepted by contemporary eye physiologists.

(6) Examinations could be partly devoted to testing students' ability to recognize cases of pseudoscience not previously considered in class and justifying their reasons for their judgments.[36]

SCIENCE EDUCATION AND THE SCIENTIFIC METHOD

One reason for studying pseudoscience in science education is that it seems to be an excellent way of showing the misuse and abuse of the scientific method. Studying pseudoscience thus provides an illuminating contrast to the way science should operate methodologically. However, it has been argued by Conant and other science educators that there is no such thing as the scientific method, there are only scientific methods.[37] Indeed, this view has become so influential in science education that a student's adherence to it is taken as evidence that the student understands science.

This view seems to be based upon a confusion between the *techniques of science* and the *method of science*.[38] Clearly different techniques of observation and hypothesis testing are used in different sciences. An anthropologist may use the method of participant observation, an astronomer may use a telescope, a sociologist may use statistical analysis, a biologist may use a microscope, and so on.

However, these different techniques do not show that there are no important structural similarities common to all the sciences. In fact, there seem to be important structural similarities among the procedures of the anthropologist, astronomer, sociologist, biologist, and so on. All these scientists test their hypotheses by deducing consequences from them, together with auxiliary hypotheses. Moreover, the general criteria of confirmation or refutation of a hypothesis do not differ in the various sciences. It is these general properties and considerations – the ones discussed above and others to be discussed in later chapters – which are usually referred to when people speak of *the* scientific method.

The ability to see the similarities among the sciences and the differences between science and pseudoscience should be considered evidence of a person's understanding of science. Thus I would maintain, against Conant and other science educators, that one evidence of a person's understanding of science is his belief that there *is* such a thing as the scientific method – at least when this is understood as it is here, which is, I believe, the way philosophers normally understand it.

Now it should be stressed that the differences between the confirmation approach and the refutation approach considered above have nothing to do with the question of whether there are several scientific methods or one scientific method. The differences are over how the scientific method is to be characterized. Both assume that there is one scientific method. In particular, both assume that the astronomer, anthropologist, biologist, and so on – whatever techniques they might use in their work – test their hypotheses by deducing consequences from them and by comparing these consequences with the evidence; both approaches assume that this testing procedure is one central aspect of the scientific method. We have found no reason so far to suppose that this supposition is mistaken.

•

CHAPTER TWO

Explanation

•

As we have seen from the Introduction, teachers like Mr. Smith are sometimes unable to explain certain things to their classes—not because of incompetence, but because science has yet to discover an adequate answer to the question. As we have suggested, this indicates that scientific explanations must meet certain standards and that in certain cases no existing account can meet the requisite standards. The question arises of whether the same sort of standards always apply.

Further reflection suggests an ambiguity in the term "explanation." Sometimes one speaks of a person explaining (or failing to explain) something to someone, e.g., Mr. Smith failing to explain the extinction of the dinosaurs to his class. But at other times one speaks of some theory or law or account explaining (or failing to explain) something. A certain theory, for example, may fail to explain the extinction of the dinosaurs.

But reflection suggests even more ambiguities. Suppose a person whom we will call Dr. Marion is working in a space research laboratory. He is asked to engage in a controversial research project, but he declines and explains why to his superiors. He requests that he stay in his present research job. Offhand, it seems that Marion's explaining why he did not want to engage in the project is a rather different sort of thing from Mr. Smith's explaining something to his class. Smith is trying to cite *causes* of why something happened; Marion is citing *reasons* for his refusal, he is justifying his action. To compound the ambiguity, not only can Dr. Marion be said to be explaining something to someone, although he is not, like Mr. Smith, interested in citing causes, but Marion's reasons might themselves be said to explain (justify) something. However, this again seems quite different from the explaining done by Mr. Smith's account in terms of causes.

Clearly the first job in considering explanation in science is to untangle some of these ambiguities. After we have achieved some clarity on the various things "explanation" refers to, two formal notions of explanation that have played an important role in recent philosophical discussions of explanation will be discussed, namely the deductive-nomological model and the statistical-probabilistic model. Once these models are understood, the question of whether explanation is the central notion in science can be discussed. It shall be argued that explanation is not the central notion in science but that understanding is, and it will be indicated how the two formal models can be generalized to cover such understanding. The relation of prediction to explanation and understanding will also be discussed.

With this general theoretical background, the role of explanation in science education can be considered. The relevance of our discussion to textbook writers, researchers in science education, science teachers, and curriculum planners will then be shown.

In the last part of the chapter, one type of explanation found in biology will be considered in detail: functional explanation. First the notion of function will be analyzed and then functional explanation will be critically considered. Finally the role of function in biological education will be discussed.

THE AMBIGUITY OF "EXPLANATION"

The word "explanation" and its cognates are used to refer to various things in science. We will distinguish two strands. First, "explanation" may refer—among other things—to (1) clarification of words or phrases, (2) justification of beliefs or actions, (3) causal accounts of events, states, or processes, (4) theoretical derivations of laws, (5) functional accounts of organs or institutions. Secondly, and cutting across these distinctions, "explanation" may refer to (1') the actions of people who are engaged in explaining something to someone, (2') the discourse of such people, (3') the success of those who are engaged in seeking explanations, (4') the discourse of such people, (5') certain linguistic expressions which stand in certain semantical, epistemological, and logical relations to the events, states, laws, words, etc., being explained.

(1) A scientist may explain some expression, such as "tidal wave," i.e., he may *clarify* what "tidal wave" means in some particular context or how it is normally used in scientific contexts. This may involve giving a definition, but it need not, since giving a definition is not the only way to clarify meaning. (Definition will be discussed in Chapter Three.) However, it is important to realize that an explanation in this sense does not explain the *phenomenon* of tidal waves but only the phrase "tidal wave." One might understand the phrase yet not understand the phenomenon.

(2) A scientist may explain some belief or action, e.g., his acceptance of a theory rejected by his colleagues or his refusal to engage in a research project that is popular among his fellow scientists. That is, he may *justify* his belief in the theory or his refusal to engage in research. Normally justification is not called for unless there is some apparent discrepancy between a belief or action and what is commonly accepted. In justifying his belief or action, the scientist is not giving a causal account, nor is he clarifying the meaning of terms. He is, rather, citing evidence or reasons that purport to make his belief or action reasonable, or at least not unreasonable.

(3) A scientist may explain some event, state, or process, such as the extinction of the dinosaurs; he may give a causal account of dinosaurs becoming extinct. In doing so, he would be neither explaining the meaning of the word "dinosaurs" nor justifying his belief that dinosaurs became extinct, but rather citing factors he takes to be causally responsible for the extinction of dinosaurs, such as changes in temperature or vegetation.

(4) A scientist may explain why some law holds. Here causal factors would not be cited, for presumably laws are not caused. Rather, some theory would be cited to explain the law. Thus Kepler's laws are explained by Newton's theory, from which close approximations to the laws can be deduced.

(5) A scientist may explain the operation of something, e.g., the heart or a social institution in society, by explaining its function. (Whether in fact the operation of a thing is explained by its function is a question we will consider later.)

In each of the five types of explanation discussed above there is still an ambiguity.[1]

(1') By "explanation" we could mean the *activity* of someone explaining something to someone. Thus:

> (a) Smith explained "tidal wave" to Jones. (Clarifying the meaning of a term for someone.)

- (b) Marion explained his refusal to engage in the research project to Jones. (Justifying an action to someone.)
- (c) Smith explained the extinction of the dinosaurs to Jones. (Giving a causal account of an event to someone.)

(2') "Explanation" can also refer to the *discourse* used in the activity of explaining something to someone. For example:

- (a) "You see, Smith, 'tidal wave' in its ordinary use is roughly equivalent to . . ."
- (b) "Jones, to be frank with you, I refuse to engage in the project for basically the following reasons . . ."
- (c) "The causal factors responsible for the extinction of the large reptiles of antiquity — commonly known as dinosaurs — are . . ."

(3') "Explanation" can also refer to the success of certain research, deliberation, or study:

- (a) Smith explained "tidal wave," i.e., he successfully clarified the phrase.
- (b) Marion explained his refusal to work on the project, i.e., he successfully justified his action.
- (c) Smith explained the decline of the dinosaurs, i.e., he produced a correct causal account.

It is obvious, of course, that the activity of explaining something to someone is quite different from the success of research and deliberation. The activity of explaining something to someone is analogous to certain activities of a teacher, and indeed, explaining something to someone may be most common in a pedagogical setting. The success of research and deliberation may well be useful to a person engaged in explaining something to someone, but it should not be confused with that activity.

(4') "Explanation" can also refer to the *discourse* used to state the results of research and deliberation. For example:

- (a) "The meaning of 'tidal wave' among the world's leading geographers is . . ." (Entry in Smith's notebook.)
- (b) "The project is a methodological disaster. I conclude that there are at least five assumptions that are . . ." (Entry in Marion's diary.)
- (c) "Our data suggest that the decline of the large herbivorous dinosaurs, for example, the brontosaurus and brachiosaurus, were the result of . . ." (Sentence in research report.)

It should be noted that the discourse used to state the conclusions of research, e.g., what a scientist puts in his research report, may be quite different from the discourse used in explaining something to someone. Pragmatic factors enter into both cases in important ways that could well make the discourse diverge: for example, the discourse of the scientist explaining the decline of the dinosaurs to college freshmen may be different from the discourse he used in his research report. The former discourse may be free from technical terms and full of colorful metaphors and elaborations while the latter discourse may be technical, concise, and abbreviated.

(5′) "Explanation" may also refer to linguistic expressions which stand in certain logical, semantic, and epistemological relations with the word, action, event, state, or law being explained.

Consider, for example, explanation as justification. Someone might argue that formally an action, A, is justified if and only if a sentence saying that A is permissible stands in relation R, a deductive relation, with a set of principles of action and factual premises, P, which meet certain epistemological standards, e.g., are themselves justified. For example, Marion's explanation (justification) for refusing to engage in a certain research project could be construed formally as follows. Let P be the following set of principles and premises:

(1) If a project, X, is based upon mistaken assumptions and these assumptions will probably adversely affect the results of the project, and if Y is expected but not required to engage in X and there are no overriding considerations, then it is permissible for Y to refrain from engaging in X.

(2) The project is based upon mistaken methodological assumptions.

(3) These assumptions will probably affect the outcome adversely.

(4) Marion is expected but not required to engage in the project.

(5) There are no overriding considerations.

Let S be:

(6) It is permissible for Marion to refrain from engaging in the project.

Now (6) is entailed by (1) through (5). Let us suppose that (1) through (5) are themselves justified. Then, given our assumptions about justification, the above argument provides a formal justification for Marion's refusal to engage in the project.

Whether this is an accurate account of a formal notion of the justification of an action we need not decide here. What is important for our purpose is to realize what this account does. It formalizes and makes explicit the notion of justification independently of the sort of pragmatic considerations that might enter into the activity of explaining (in the sense of justifying) some-thing to someone. This does not mean that in actual practice justi-fication conforms exactly to these criteria; pragmatic considera-tions may make exact fulfillment of them out of the question.

For example, Marion may find it much too tedious and time-consuming to state all the premises in *P* in his actual justifying ac-tivity. Some of these premises may indeed be tacitly assumed in the actual context of justification. Moreover, if Marion is justifying his action to his five-year-old daughter, he may have to simplify the account somewhat, choosing words and concepts which she can understand, but which do not do full justice to his actual ratio-nale. In short, Marion may have to sacrifice complete accuracy for other, practical considerations.

These modifications in actual practice do not mean that the formal notion of justification is without use. It may be used as a standard for evaluating the *formal and epistemological aspects* of the actual activity and discourse of explanation as justification. Prag-matic considerations would also have to be used in a complete evaluation of this activity and discourse.

A similar formal notion of explanation is needed for the other types of explanation we have been discussing. A formal no-tion of causal explanation, for instance, would specify the logical and epistemological relations between linguistic expressions, e.g., sentences, and the events, states, or processes being explained. Such a formal notion would be helpful in at least three respects. First, it would enable one to evaluate the discourse used in a caus-al explanation of something to someone else, such as Jones' dis-course in explaining the decline of the dinosaurs to Smith, on purely logical and epistemological grounds independent of prag-matic considerations. Of course, pragmatic criteria would also be needed for a complete evaluation of Jones' discourse. It may be faultless logically and epistemologically, and yet be too technical for the age and background of Smith.

Secondly, the formal notion of causal explanation would set the logical and epistemological standards for the results of re-search. Put a different way, pragmatic considerations aside, the explanatory researcher should aim at producing a set of linguistic expressions meeting certain logical and epistemological criteria. Thirdly, the formal notion of explanation would provide logical and epistemological standards for evaluating the discourse of the researcher. It would answer the question: Pragmatic considera-tions aside, does the discourse of the researcher when he states his results provide a causal explanation, i.e., does it consist of sen-tences meeting certain logical and epistemological requirements?

So far we have distinguished various types of explanations. Some of these are more relevant than others for our purposes. In particular, the formal notion of causal explanation is quite relevant for recent discussions in the philosophy of science. In a moment we shall consider an attempt to formulate a formal notion of causal explanation. Secondly, in science teaching the notion of explaining something to someone is of crucial importance, and later in the chapter the role of explaining something to someone in science education will be considered. Thirdly, explaining the operation of an organ in an organism in terms of the organ's function in the organism—so-called functional explanation—is important in biological science and this sort of explanation will be considered toward the end of the chapter.

THE DEDUCTIVE–NOMOLOGICAL MODEL OF EXPLANATION

One attempt to formulate a formal notion of causal explanation is the so-called *deductive–nomological* (D–N) model of explanation. This model has played an extremely important role in recent philosophical discussions of explanation in science. It has been expounded by many well-known philosophers of science, such as Popper,[2] Hempel,[3] and Nagel,[4] and discussion of this model by way of either criticism[5] or defense[6] has dominated recent philosophical literature on the topic of explanation in science.

Stated informally, the model is this: A causal explanation of some event is achieved when that event is subsumed under some causal law. Thus someone might ask why a particular substance conducts electricity. The answer might be that the substance in question is copper and that all copper conducts electricity. This subsumption constitutes the explanation of the phenomenon in question. Again, someone might ask why a rod lengthened. The answer might be that the rod is made of copper and that the rod was heated and copper expands when it is heated. Again the phenomenon to be explained is being brought under a causal law.

Put in the form of an argument, the two explanations would look like this:

(1) All copper conducts electricity.
 This substance is copper.

∴ This substance conducts electricity.

(2) All copper expands when heated.
 This rod is made of copper.

∴ This rod expands.

It should be noted that both of these explanatory arguments have the same general characteristics. At least one of the premises is a causal law; all the others are statements describing particular conditions that hold in a given situation. The conclusion which describes the event to be explained follows deductively from the premises.

Thus the general form of explanation, according to the D – N model, is this: Given a certain set of causal laws and statements of what have been called initial conditions, a statement describing the event to be explained follows. Put in a diagrammatic way, a D – N explanation would look like this:

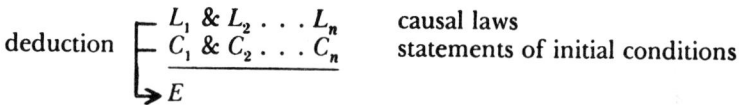

$$\text{deduction} \begin{bmatrix} L_1 \ \& \ L_2 \ . \ . \ . \ L_n & \text{causal laws} \\ C_1 \ \& \ C_2 \ . \ . \ . \ C_n & \text{statements of initial conditions} \\ \hline \rightarrow E & \end{bmatrix}$$

The laws and the sentences stating the initial conditions have to meet certain logical and epistemological requirements:

R_1 All the laws and initial conditions have to be essential for the deduction.

R_2 All the sentences have to be testable.

R_3 E must logically follow from the statement of initial conditions and laws.

There has been some disagreement over a fourth requirement. Some philosophers have argued:

R_4 All the sentences in the explanation must be true.

Others have argued:

R'_4 All the sentences in the explanation must be well confirmed relative to available evidence.

It should be clear that R_4 and R'_4 are logically independent requirements—neither requirement entails the other. Some sentences might be true without the available evidence confirming them. On the other hand, some sentences might be well confirmed by the available evidence and yet be false. In the first case, a scientist would have no justification for supposing he had a true explanation; in the second case the scientist would be justified in supposing that the explanation was true although the explanation was not true. In any case, it appears that there is no real disagreement here since two different formal notions of causal explanation are at issue. R_1, R_2, R_3, R_4 specify the requirements for a *true causal explanation;* R_1, R_2, R_3, R'_4 specify the requirements for a *justified causal explanation.*

Let us consider one of the examples already given in the light of these requirements, namely, the explanation of why the rod lengthened as expressed in argument (2) above. The first requirement, R_1, is certainly fulfilled. The law and statements of initial conditions are essential for the deduction; none of these statements can be omitted if the conclusion is to follow. The second requirement, R_2, holds also since these sentences are testable. R_3 holds, for the conclusion "This rod expands" logically follows from the premises. Furthermore, R'_4 holds, for in the light of the available evidence the law and initial conditions are well confirmed. So argument (2) is a justified causal explanation of why the rod expanded. Moreover, according to confirmation theorists, R'_4 gives us good reason to suppose that R_4 holds. Hence, in their view we have good reason to suppose that this is a true causal explanation.

Again it should be stressed that pragmatic considerations may make it unnecessary or undesirable for the actual discourse of one who explains something causally to someone to conform exactly to the D – N model. For example, Jones may explain the lengthening of the rod to Smith by saying, "You see, the rod was heated." This may be all that is necessary or desirable to say to Smith. Nevertheless, Jones' discourse would be explanatory from a logical and epistemological point of view only because other things were tacitly assumed, namely, that the rod was made out of copper, that the law holds, and that the description of the event explained can be deduced from the law and the statement of initial conditions. The D – N model thus clarifies the tacit assumptions that must be made in an explanatory activity if its discourse is to be logically and epistemologically adequate.

The above point has often been misunderstood in criticism of the D – N model. Some philosophers of science have seemed to assume that, because the actual explanatory discourse of scientists does not conform to the requirement of the model, the model is incorrect. But the D – N model does not purport to reflect the actual explanatory discourse of the scientist, any more than the forms of arguments in logic books purport to reflect actual discourse. The D – N model purports to specify epistemological and logical requirements for explanatory discourse abstracted from practical considerations, just as the forms of arguments in logic books purport to formulate criteria for valid arguments abstracted from practical considerations.

To be sure, the requirements specified by the model might be wrong, and arguments might be offered which show that one or more of these requirements should be changed. However, merely showing that scientists do not put their explanatory discourse into D–N form, or that doing so would be inconvenient for scientists, is as irrelevant to showing that the D–N model is incorrect as showing that people do not put their arguments into syllogisms, and that it would be inconvenient for them to do so, is irrelevant to showing that the requirements in a logic textbook are incorrect.

In any case, one of the major advantages claimed by advocates of the D–N model is its ability to indicate the logical and epistemological problems in explanations. Consider, for example, the following piece of explanatory discourse:

> The causal factor responsible for the extinction of the large reptiles of antiquity—commonly known as dinosaurs—was the change in vegetation brought about by a change in climate. The plant-eating dinosaurs could not eat the tougher vegetation and died out. The flesh-eating dinosaurs who preyed on the plant-eating ones perished in turn.

What assumptions are being made by the speaker and what factual support is there for them? Perhaps these assumptions can be spelled out and their backing elaborated. If so, we might have a full-fledged D–N explanation. However, if they cannot be, this discourse will simply be an outline or a sketch of a causal explanation—what has been sometimes called an *explanation sketch.*[7] That the discourse as it stands needs to be filled in in accordance with the D–N model is made clear by the following:

SKETCH OF D–N EXPLANATION I

	?
Laws assumed.	
Statement of initial conditions apparently assumed.	C_1 Some dinosaurs are plant eaters.
	C_2 During a certain period of time, $t_1 - t_2$, plants in the dinosaurs environment become tougher.
Other statements of initial conditions.	?

∴ Plant-eating dinosaurs died off during period $t_1 - t_2$.

SKETCH OF D–N EXPLANATION II

Laws assumed.	?
Statement of initial conditions apparently assumed.	C_1' Some dinosaurs were flesh eaters.
	C_2' The flesh-eating dinosaurs preyed on the plant-eating dinosaurs.
	C_3' Plant-eating dinosaurs died off during period $t_1 - t_2$.
	C_4' Flesh-eating dinosaurs could find no other animals to prey on that would sustain them.
Other statements of initial conditions.	?

∴ Flesh-eating dinosaurs died off during period $t_1 - t_2$.

Putting the explanatory discourse into D–N form exposes the logical gaps and makes explicit what assumptions are being made and what others may have to be made. One can begin to see possible weak spots in the explanatory argument. For example, is C_3 true? If so, why weren't the plants too tough for other plant eaters to eat? It is clear that not all the plant-eating animals died out during this period. Is C_4' true? (This condition surely must be assumed, for unless it is there would seem to be no reason why the flesh-eating dinosaurs could not have survived on nondinosaurs.) What is the supporting evidence? If there is none, this is a weak spot in the sketch. What laws are assumed by the explanation? What is the supporting evidence for these alleged laws?

Bringing questions like these to the fore has at least two values. First, it provides the scientific investigator with some guidelines for research in filling in the details of the sketch and producing a more complete explanatory argument. Secondly, it provides the science teacher or student of science with insight and understanding into the gaps in our scientific knowledge. We shall return to this point later when we consider explanation in the context of science education.

THE STATISTICAL–PROBABILISTIC (S–P) MODEL OF EXPLANATION

So far we have assumed that the explanatory discourse should be evaluated in terms of the D–N model. Such a model assumes that general laws are necessary for a full-fledged explanation. Thus we have assumed that the laws presupposed in the explanation would have the form "All A and B." For example, perhaps a rough statement of a law presupposed in I above is:

> L_1 All plant-eating land animals die when they no longer are able to eat the plants within a radius of two thousand miles from where they live.

However, this is not the only possible reconstruction of explanatory discourse. Perhaps instead of general laws being assumed, statistical laws are assumed. Such a reconstruction has certain advantages. Perhaps L_1 as it stands is false. Perhaps some land animals might be capable of traveling long distances to new environments where they can eat the vegetation. A more plausible assumption might be:

> L_2 Most plant-eating land animals die when they no longer are able to eat the plants within a radius of two thousand miles from where they live.

Now such a statistical law could not explain—even when combined with appropriate initial conditions—why all dinosaurs died out; at most it would explain why most of them did. However, this incompleteness may suggest that more laws—either general or statistical—are needed for a complete explanation.

In any case, the consideration of statistical laws suggests a different formal model of explanation. Let us call this model the *statistical–probabilistic* (S–P) model.[8] This model is like the D–N model except for two things: (1) The laws in the premises of an explanatory argument are statistical laws rather than general laws. An example of such a law would be L_2 above. Such laws might be stated in a precise quantitative form, e.g., 90 percent of A's are B, or in a less precise way, e.g., Most A's are B, or Nearly all A's are B, or The proportion of A's that are B is close to 1, or Any A has a good chance of being B. (2) The relation between the premises and the sentences describing the event, state, or process to be explained is probabilistic rather than deductive. Consider, for example, the following S–P explanation. We will assume that the premises of the argument are true.

> Nearly all people having streptococcal infections who are treated with penicillin recover.
> Jones had a streptococcal infection and was treated with penicillin.

> ∴ Jones recovered.

Now the conclusion cannot be logically deduced from the premises as in a D–N explanation. For the premises are true and yet the conclusion could be false. However, in a valid deductive argument if the premises are true, the conclusion must be true also. Nevertheless, the conclusion is probable relative to these premises. From this example we can abstract the general form of S–P explanations and diagram this form as follows:

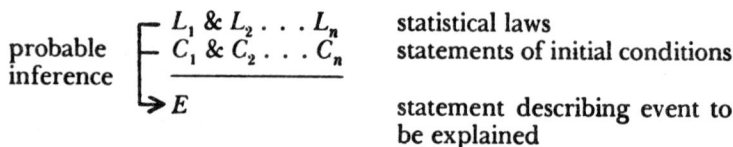

$$
\text{probable inference}
\begin{cases}
L_1 \ \& \ L_2 \ldots L_n & \text{statistical laws} \\
C_1 \ \& \ C_2 \ldots C_n & \text{statements of initial conditions} \\
\hline
E & \text{statement describing event to be explained}
\end{cases}
$$

Another example of an explanation with this form can be found in evolutionary theory.[9]

Suppose one is puzzled by the following phenomena: light-colored moths have not survived in a particular environment. Suppose that one learns that the environment is sooty and that moth-eating birds inhabit this environment. Now suppose the following sentences describe the event to be explained:

E_1 = Dark-colored moths have survived.
E_2 = Light-colored moths have not survived.

Now consider the following statistical laws derived from evolutionary theory:

L_3 In any environment, if any organisms possess advantageous characteristics lacked by other organisms of the environment, those organisms will have a good chance of surviving.

L_4 In any environment, if any organisms lack advantageous characteristics possessed by other organisms of the environment, those organisms will have very little chance of surviving.

Such laws combined with certain statements of initial conditions make E_1 and E_2 very probable relative to these laws and initial conditions. What sorts of initial conditions would these be? It seems clear that one would have to specify some property possessed by dark-colored moths that gives them an advantage. This presumably would be their dark color, which makes it more difficult for birds to find and eat them in this environment. Thus:

C_1 Dark-colored moths because of their color possess an

advantageous characteristic in this environment which light-colored moths lack.

C_2 Light-colored moths because of their color lack an advantageous characteristic possessed by the dark-colored moths.

L_3 and L_4 plus C_1 and C_2 make E_1 and E_2 probable. Thus L_3 and L_4 and C_1 and C_2 provide a statistical–probabilistic explanation of the biological phenomena in question.

It should be noted that the phenomena described by E_1 and E_2 could have been predicted had we known the initial conditions prior to their occurrence. Thus it is not true, as has sometimes been alleged, that evolutionary theory allows for explanation but not for prediction. Predictions are possible if the initial conditions are known. Of course, it is often difficult to know the initial conditions, but this is a reflection on us rather than on evolutionary theory. (We will discuss the relation between prediction and explanation later in the chapter.)

It may be objected that the above explanation does not meet R_2, that L_3 and L_4 are in fact tautologies and thus cannot be tested. Thus it may be said that any organism which possesses advantageous characteristics would by definition have a good chance of surviving, since "advantageous" in this context only means that it aids survival. Hence L_3 and L_4 are true by definition.

However, this argument is mistaken. To say that an organism, O_1, possesses advantageous characteristic C lacked by another organism, O_2, in environment E is to say:

(1) In E there is some condition N which must be satisfied by O_1 and O_2 or else O_1 and O_2 will die off and not reproduce themselves.
(2) C satisfies N or causes N to be satisfied.
(3) Either O_2 possesses no means of satisfying A or O_2 possesses less effective means of satisfying N in E than O_1.

N might be avoiding enemies, or seeing food, or getting water; and C might be running fast, or not being visible. Let us understand the phrases "have a good chance of surviving" and "have very little chance of surviving" to refer to a long-range frequency of survivals. Thus to say that O_1 has a better chance of survival than O_2 is just to say that in the long run the frequency of O_1's surviving is greater than the frequency of O_2's surviving.

Construed in this way—and this seems to be the way biologists actually construe L_3 and L_4—L_3 and L_4 are not tautologies. There would be nothing self-contradictory about an organism that had an advantageous characteristic yet did not survive, or about an organism that lacked such a characteristic yet did survive. Thus we see that L_1 and L_2 are not true by definition.

It is possible in branches of science in which statistical laws are used that these statistical laws will someday be replaced by general laws. Thus the statistical laws of evolutionary theory may someday be replaced by general laws. Such a replacement might involve, for example, discovering some additional property of organisms or of their environments which, when combined with the organisms' advantageous characteristics, would provide a nomologically sufficient condition for survival. The general form of a replacement would be this:

(1) Original statistical law:
   For every x, if x has A, then with frequency F x has B.
(2) Replaced by general law:
   For every x, if x has A and P, then x has B.

The replacement would turn on finding a suitable property P. Thus the use of statistical laws in science is logically compatible with the existence and eventual discovery of such a property P, and with the existence and eventual discovery of general laws. Whether such general laws do exist and whether, if they do, they will ever be discovered is another question.

SCIENTIFIC UNDERSTANDING

We have discussed the D–N model and the S–P model in reference to causal explanation. However, they have a broader significance for science. These models help illuminate certain noncausal explanations in science. Moreover, they provide insight into other sorts of inferences in science, in particular, into scientific understanding. Scientific understanding goes beyond explanation—causal or otherwise—although it includes explanation. To understand an empirical phenomenon, we shall argue, is to see how it fits into the nomological nexus—the network of laws, general and statistical, that connect the phenomenon at issue with other phenomena past, present, and future. However, as we shall see, only some of these lawful relations would be considered explanatory.

First let us consider some applications of the D–N model to noncausal explanations. Similar applications could easily be found for the S–P model.

Broadly interpreted, the D–N model provides an account of the explanation of laws by means of theory. For example, given certain laws of Newtonian mechanics and statements of certain initial conditions, close approximations of Kepler's laws can be derived. These would not provide a causal explanation of Kepler's laws; they would, however, explain these laws.

The D–N model provides a specification of the logical and epistemological requirements for explanations of events, states, or processes which would not normally be called causal.[10] Thus, given a law governing a simple pendulum

$$t = 2 \pi \sqrt{l/g}$$

and given the knowledge that the arm of some particular pendulum is one hundred centimeters (as well as the value of g, the acceleration for free fall), we can deduce the period of the pendulum as two seconds. This deduction explains the two-second period; the explanation is hardly a causal explanation, however, if for no other reason than that the length of the pendulum and the period were simultaneous and one does not normally consider simultaneous events and states as causing one another.

The existence of these noncausal explanations in science is important. It has often been noted that references to causes tend to disappear as science becomes more mature and theoretical, that few explanations in mature sciences are causal. The laws used in explanations in mature science tend to be precise, mathematical, functional statements having no obvious causal import. Causal talk thus seems characteristic of the more immature sciences and the applied sciences, such as medicine.[11]

Perhaps another reason for the decline of causal talk in science is the obscurity of the notion of causality itself. Attempts to replace the notion of cause by clearer notions have proved unsuccessful. Consider the following expression:

(1) A is the cause of B.

Some philosophers have suggested that (1) is equivalent to

(2) A is a sufficient condition for B and A occurs prior to B.

However, this analysis is too broad. Consider a case in which C causes B as well as A (C is the common cause of A and B), A is prior to B, and A is a sufficient condition for B. This is an instance of (2) but not of (1).

Whatever the reason for the decline of the importance of causal explanations in theoretical science and the existence of other types of explanation that fit the D–N model, it is an open question whether all scientific explanations fit either the D–N model or the S–P model. It is possible that these models specify the logical and epistemological requirements for a large class of scientific explanations but not for all. We will return to this problem again when we consider the explanatory status of functional statements.

Secondly, let us examine how the D–N model goes beyond explanation—causal or otherwise.[12]

Suppose that a geologist describes an earthquake in 1805 (event e_1) as E. Now from some law L and initial condition C describing some geological event in 1808 (event e_2), he deduces E. Such a deduction would not explain e_1 since e_1 is before e_2. Nevertheless, such an inference is in accord with the D–N model. In fact the D–N model does not require any particular temporal relation between e_1 and e_2 at all. However, explanation can only occur when e_2 is either before or simultaneous with e_1.

Moreover, even when the conditions specified by C occur before or simultaneously with the event to be explained, the D–N model allows arguments which are not explanatory. Consider another example. There is a flagpole twenty-five feet high. This flagpole's height is described by sentence E. Given some laws L of optics and sentences C describing the angle of the sun and the length of the shadow cast by the flagpole, the height of the flagpole can be deduced. Thus L and C entail E. However, it would be implausible to say that the height of the flagpole was explained; certainly the question of why the flagpole was twenty-five feet high was not answered.[13] Nevertheless, the inference fits the D–N model.

Notice that, despite the fact that in these two cases explanations have not been given of the phenomena in question, the earthquake and the flagpole's height, our understanding of the phenomena has increased. The phenomenon has been connected with other phenomena by means of nomological relations specified by laws. In the earthquake case, although we don't know why the earthquake occurred, we know how the earthquake is connected with some geological event that came after it, and in knowing this we have gained some understanding of the earthquake. The earthquake of 1805 is no longer seen as an isolated event but as a part of a nomological nexus—lawfully related to events prior to, simultaneous with, and after it in geological and cultural history. To simply know the earthquake's connection with a later geological event may not provide much understanding of the earthquake, but it does provide some knowledge of other connections with events later than, earlier than, and simultaneous with earthquakes. Of course, knowledge of a theory in which these connections are unified will give an even fuller understanding.

In the flagpole example the situation is similar. We do not know why the flagpole is twenty-five feet high by deducing E from L and C. However, we do know how the flagpole's height is lawfully connected with other phenomena, i.e., its shadow given the angle of the sun. And this knowledge does provide some understanding of the flagpole's height. Again, we do not see the phenomenon as an isolated one, but rather as a phenomenon connected with other phenomena by nomological relations. To be sure, such understanding is limited and more lawful connections would have to be known for any fuller understanding of the flagpole's height: its connections with the builder who erected it, with traditional conceptions of flagpole heights, with available materials, with later conceptions of patriotic symbols, with the landscape that surrounds it, and so on. Furthermore, these various connections would have to be seen as unified or related in terms of theory or theories to have the sort of understanding that is characteristic of science. Nevertheless, although other connections would have to be known and a theory would have to be available to unify these connections for a full understanding, some understanding is provided by showing the flagpole's connections with its shadow and the angle of the sun, i.e., by the D–N inference specified above.

We have suggested above that, as science matures, causal explanations become less important. It also seems to be true that, as science matures, explanation—causal or otherwise—becomes less important. The major aim of mature theoretical science is not to explain phenomena, but to achieve a scientific understanding of them. Scientific explanation is only a special case of such understanding. The first aim of mature theoretical science is to discover lawful relations among empirical phenomena—relations that enable us to make inferences to events past, present, and future. Only some of these inferences would be regarded as explanatory. The second aim of theoretical science is to provide theories that unify these lawful relations. The laws of science increase our understanding of phenomena by showing the lawful connections among phenomena; the theories of science deepen our understanding of these phenomena by showing us how these laws themselves are connected by some unifying idea or principle.

Scientific understanding, as we outline it here, is understanding of the phenomena of the world. But there are other kinds of understanding. For example, one can understand science, that discipline that seeks to achieve a certain kind of understanding of the world. To be sure, one's understanding of science could be a scientific understanding. In this case one might regard science as a particular cultural phenomenon. Sociologists of science might try to discover lawful relations connecting various parts of this discipline to each other and to other cultural phenomena and environmental factors. However, sometimes when one speaks about understanding science something different is meant. In these cases, science is not regarded as one cultural phenomenon among others, but as a way of studying all phenomena, as an approach or method or point of view. In these cases something different from *scientific* understanding is involved. We shall give a more general analysis of understanding, and of understanding science in particular, in the last chapter. It is sufficient to say now that scientific understanding as we consider it here consists of knowledge of certain lawful relations and of theories which unify these relations. As we shall see, understanding does not always consist of this sort of knowledge.

How do the D–N and S–P models fit in with scientific understanding as characterized in the above way? As we have seen, the D–N and S–P models apply to inferences made according to laws—general or statistical. Given description E of event e_1, a scientist using law L and initial condition C infers E. Only some of these inferences, as we have seen, would be considered explanatory. However, these are the sort of inferences that provide scientific understanding to a certain degree. As we have seen, scientific understanding of event e, under description E, consists partly of knowing the lawful connections between e and other events before it, after it, and simultaneous with it. Knowledge of these lawful connections is formalized by the D–N and S–P models, since to know, for example, that event e_1 and event e_2 are lawfully connected in a particular way may be just to know that, given E, one can infer E from law L and initial condition C which describes event e_2. Furthermore, the two models provide a formalized way of talking about a theory's unification of laws. For to say that theory T unifies laws L_1 & L_2 . . . L_n is usually to say that, given L_1 & L_2 . . . L_n, one can infer L_1 & L_2 . . . L_n from T, given certain initial conditions. However, this account is simply a generalized version of two models; instead of events being explained, we have laws being unified—nevertheless, the formal structure remains the same.

We have argued that explanation is not the major aim of theoretical science, but rather that scientific understanding is the major aim, and that explanation is a special case of this understanding. Furthermore, we have maintained that the D–N and S–P models, considered in a generalized form, are relevant in explicating this understanding. However, one might well ask what role prediction has in science and what relation prediction has to the formal models considered above. For example, what is the relation between prediction and explanation according to the D–N model?

According to some advocates of the D–N model, prediction and explanation have the same form; there is only a temporal difference between them. In prediction, sentence E is a description of some event (or state or process) that has not yet occurred; in explanation, E is a description of an event (or state or process) that has occurred.

However, the formal identity of prediction and explanation can only be maintained if the notion of prediction is restricted. Prediction by means of general laws and initial conditions would of course have the same form as D–N explanations. But some predictions are made without recourse to any law, general or otherwise. This is the case with the prediction of a fortune-teller.

In any case, predictions which are formally identical with D–N explanations are only a special case of a more general pattern of inference found in theoretical science. (In this respect also, D–N predictions are like D–N explanations.) Another pattern of inference specified by the D–N model is this: Given law L and sentence C describing event e_2, one deduces E which describes event e_1.

However, one would normally say a prediction was made only when e_2 occurred before e_1. (Indeed, one might even restrict predictions to inferences in which e_2 occurred before e_1 *and* in which e_1 occurred after the prediction was made.) In any case, there are two other possibilities:

(1) e_2 occurs simultaneously with e_1.
(2) e_1 occurs before e_2.

Theoretical science is not only interested in making predictions; it is interested in making inferences in which (1) and (2) hold. For example, given some law L and statement of initial condition C describing some solar phenomenon in 1932, a physicist might infer some solar disturbance in 1930. This would not normally be called a prediction, although it has the same form as a D–N prediction.

We have seen that explanation is not the goal of theoretical science. Even less is prediction the major goal of science. In the first place, accurate predictions might be made by fortune-tellers and soothsayers. However good these predictions are, making *these* kinds of predictions is not the aim of science. Such predictions are not made by means of laws, and the predictions of science are. Science is only interested in these kinds of predictions as phenomena to understand. Secondly, even if the predictions are made by laws, we have seen that they are not the only sorts of inferences that are important to science. Predictions are just a special case of a more general type of inference. Like causal explanation, which they are associated with, they are of primary importance in applied science, e.g., medicine, but not theoretical science. Thirdly, even when inferences of the general type are used, science is interested in unifying these inferences by means of theories.

As we shall see in the last chapter, the secondary importance of prediction in theoretical science is important since some science educators have argued that the major goal of science education is leading children to predict. We shall consider this goal of science education critically in the last chapter.

EXPLANATION AND SCIENCE EDUCATION

Although the major goal of science is not explanation, explanation is still an important concept for science and for science education. In this section we will consider the role of explanation in science education. We will show the relevance of our philosophical analysis of explanation for the science textbook writer, the researcher in science education, the science teacher, and the science curriculum planner.

SCIENCE TEXTBOOK WRITERS

The fruitfulness of the D–N and S–P models of explanation in science education has been too long neglected. The use of these models would illuminate many of the discussions of explanation in science textbooks. Consider the following example.

As we mentioned earlier, one of the unsolved problems of science is the disappearance of the dinosaurs. That it is an unsolved problem is often mentioned in science textbooks. Thus *Biological Science: An Inquiry into Life* says,

> In late *Cretaceous* (kre-*tay*-shus) times, barely 100 million years ago, the dinosaurs reached their evolutionary peak. Many new groups, each adapted to a different way of life, developed, giving the reptiles a variety never previously attained. Then, while seemingly at the height of their development, these mighty reptiles became extinct. The reason is one of the puzzles of paleontology. Numerous theories have been advanced; not one has been generally accepted.[14]

Unfortunately, the authors of *Biological Science* do not discuss what some of these theories are or why they have not been accepted. It is not implausible to suppose that from such a discussion science students might learn a great deal about what is an acceptable scientific explanation. Indeed, there is no general discussion of what constitutes an acceptable scientific explanation in the book, despite the fact that actual scientific explanations are discussed throughout. The D–N and S–P models could have been used not only as tools for evaluating the various explanations of the disappearance of the dinosaurs, but as general frameworks in which to place the causal explanation given in the book.

E.S.C.P.'s *Investigating the Earth* says this about the disappearance of the dinosaurs:

The relatively sudden disappearance of the dinosaurs . . . took place at the end of the Cretaceous Period. Although these creatures had ruled earth for more than 140 million years, their rapid decline brought the Age of the Reptiles to an end. Why should these animals, which had so successfully adapted to such a wide variety of habitats, come to such an abrupt end at the peak of their development? Can you propose any explanation for their sudden extinction?[15]

Unfortunately, the authors of *Investigating the Earth* give students little guidance in deciding how they might evaluate an acceptable scientific explanation of the dinosaurs' disappearance. Yet it is difficult to see how students could be expected to come up with a plausible explanation without such guidance. Again the book contains no general discussion about what an acceptable scientific causal explanation would consist of. In this case also it would seem that a general account of scientific explanation, for example, a discussion of the D–N explanation, would be illuminating and helpful.

RESEARCHERS IN SCIENCE EDUCATION

As we have already seen, in one sense of the term, explanation is the activity of explaining something to someone. The teacher engages in this activity, but not all of the teacher's activity is explanatory and not all discourse used by the teacher is explanatory discourse.[16] Indeed, not all the activity that is aimed at getting a student to learn the explanation of something would be counted as explanatory activity. For example, a biology teacher takes students on a field trip. On this trip the teacher says very little and what he does say is not explanatory, e.g., "See that redwing blackbird? What is he doing?" Nevertheless, the teacher hopes that the students, from what they observe on the field trip, will be able to figure out why birds build nests. Thus there are alternative ways for students to learn explanations. One and only one of these ways is the explanatory activity of the teacher.

An important job for the researcher in science education is to determine which is the best means for acquiring explanations. This is in part an empirical question, since it involves factors such as efficiency and the lasting quality of the learning produced. In part, however, the question is a normative one, since "the best way" usually suggests a fair and humane way as well as an efficient way. The question perhaps can be put as follows: Supposing that explaining something to someone is a fair and humane way (which it seems to be in many situations) to learn the explanation of something, is it as efficient, long lasting, and so on as other fair and humane ways? The answer will undoubtedly be relative to the sort of explanation and type of student at issue. College freshmen may learn the explanation of the red shift better from the explanatory activities of a science teacher than in any other way, while high school freshmen may learn the explanation of certain animal behaviors better from a field trip in which no explanatory activity by the teacher occurs. However, this cannot be determined in any *a priori* way and empirical investigation would certainly be needed. One might teach two groups of science students different explanations by different methods. One group would learn an explanation by means of the explanatory activity of the teacher; the other group would learn the same explanation by some alternative way, e.g., going on a field trip and being presented with various clues. The two groups would be tested later to see how well they learned via the two methods. They might be tested for retention, ability to apply the explanation they learned to similar cases, and so forth.

SCIENCE TEACHERS

The above analysis of explanation should be helpful to the science educator in at least two ways: in how the science teacher teaches (method) and in what the science teacher teaches (content).

We have seen from our discussion of research on different methods of getting students to learn explanations that a teacher can approach his goal in different ways. Educational research may help the teacher decide which way he should use. But if there is no evidence available from research at the time, the science teacher must do the best he can. He should experiment with different methods in his classroom and see which works best.

However, if a science teacher decides to explain something to his class rather than use some other method to get the class to learn an explanation, our previous discussion is suggestive of how he should proceed. We have emphasized repeatedly that, in the discourse used in the activity of explaining something to someone, pragmatic considerations are important. The teacher as explainer should adapt his explanatory discourse to the age and level of the students. Complete accuracy may have to be sacrificed to increase student understanding. The general maxim for the explanatory discourse of the science teacher is to maintain as much accuracy as is compatible with assimilation and understanding of the material by the students. Thus the explanatory discourse of a good science teacher — even if it is quite accurate — will usually be different from the discourse of the scientist stating the results of his research activity. As we have seen, the science research report may be brief, cryptic, and technical, and because of this quite unintelligible to the science student. In order to make such research intelligible to his class, the science teacher may have to use analogies and examples familiar to the students, draw on the students' experiences, state the same point in several different ways, shift the focus of the discussion, and so on. All of this should be done with as few distortions of the correct scientific explanation as possible.

Some science teachers acquire this skill of explaining something to their classes in an intelligible and fairly accurate way with little effort; for other science teachers, it is hard to come by. The discourse of the teacher is either quite accurate but completely over the head of the student, or else colorful and lively but needlessly inaccurate. In any case, developing such a skill is important enough that serious training should be given to it in schools of education. Unfortunately, this is usually not done.

We have argued so far that the activity of explaining something to someone is one way that science teachers can teach explanations to their students. However, explanatory activity can also be what science teachers teach, and not merely what they engage in. In short, explanation can be part of the content of a science education course.

Students in science courses might be taught how to seek explanations. Presumably learning this skill would involve more than just reading about explanation in science, but would also involve guided practice. This practice would be closely connected with learning how to do scientific inquiry, since one goal of scientific inquiry is to produce explanations.

Students could also be taught how to explain something to someone in scientific terms. It is part of a liberal education to learn how to express oneself, to get across ideas to various people in various circumstances; unfortunately, however, this aspect of science education has been largely neglected. People trained in science are often inept in explaining something scientific to someone else.

Now it might be argued that, although this is indeed an important goal of education, it should not be a goal of *science* education, that learning the skill of giving a scientific explanation to someone should be achieved in English or speech courses. However, this contention is by no means obvious. For one thing, the English or speech teacher is usually ill equipped to judge the logical and epistemological accuracy of a student's scientific explanation. Could courses in English and speech teach students at least the pragmatic dimensions of explaining something in science to someone? There is no doubt that such courses could help. But it is doubtful that they could do the entire job, for it is doubtful that the skill of explaining something scientific to someone is a simple combination of the knowledge acquired in a science course and the literary and expository skills acquired in speech and English courses. A person who is good at giving a scientific explanation to someone must combine in a very delicate way his scientific knowledge and the contingencies of the explanatory situation. Learning how to achieve this delicate balance may well be something that is best learned in situations in which explanatory activity is critically examined both from a logical and epistemological perspective and from a pragmatic perspective. A science course in which students are taught how to explain something they learned in class to the class, to the instructor, to their parents, and so on, may provide this situation.

SCIENCE CURRICULUM PLANNERS

A science curriculum could be built around the notion of explanation. Scientific explanation could be one of the unifying themes of the curriculum. That is to say, explanations from various sciences could be studied in such a way as to bring out the similarities and differences. Consider, for example, a science curriculum built around the following explanations: Torricelli's explanation of the height of water raised by a suction pump,[17] Redi's explanation of the appearance of maggots on uncovered meat, Semmelweis' explanation of childbed fever,[18] Durkheim's explanation of suicide rates, Freud's explanation of slips of tongue,[19] Skinner's explanation of language acquisition.

In such a curriculum one might wish to use material which illustrated scientifically unacceptable explanations, as well as paradigms of acceptable science explanations, in order to show how the logical or epistemological requirements of explanation can fail to be met. For this purpose, material drawn from the literature of pseudoscience may be appropriate. For example, Bates' "explanation" of accomodation could be used. There might also be some point to using cases of explanation that are debatable scientifically, in which it is unclear whether the explanation is acceptable scientifically or not. Freud's explanation of slips of the tongue might be an example of this.

A study of various types of explanation in science could be approached by reading case studies in the history of science or original papers in science. But this is not the only way to approach the subject. A textbook might be written in which explanation in science was the major theme. Many of the aspects of scientific explanations discussed in this chapter could be covered in such a book, except at a more elementary level, and case studies from the history of science could be used to illustrate the various points. Naturally, other topics that are important in understanding science would be brought in and integrated with the discussion of explanation, e.g., observation, hypothesis testing, criteria for confirmation, and the relation between natural, formal, and social sciences.

We have suggested above that, to illuminate acceptable scientific explanation, unacceptable explanations as well as debatable cases might be considered. We mentioned Freud's explanation of slips of the tongue as a possible case of the latter. Freud's case is debatable because of the dubious scientific status of psychoanalysis. However, in other debatable cases there is no analogous problem. Most scientists do not have doubts about the scientific status of biology. Nevertheless, it is unclear whether certain statements that appear in biological science are explanatory. The question is not over the scientific nature of biology, but over the nature of explanation. In the next section we will consider the status of functional statements as explanations of biological phenomena.

FUNCTIONAL EXPLANATIONS IN BIOLOGY

Biologists often talk about a function of a certain part of an organism or of some activity of this part. Here are some typical examples:

> A function of the heart's pumping blood is to circulate the blood.
> A function of breathing is to maintain an adequate oxygen supply in the blood.
> A function of the archinephric duct is to transport sperm and urine out of the organism.

It should be noted that all of these statements have one form:

> (1) A function of X is Y.

We are interested in the following question: If we know that a function of X is Y, have we given a scientific explanation of X? Much will depend on what is meant by (1). Some philosophers have argued that (1) means:

> (2) X is a necessary condition of Y.[20]

Thus, to say that a function of the heart's pumping blood is to circulate blood is just to say that the heart's pumping blood is a necessary condition for circulating the blood.

There are two major objections to this analysis, however. First, it is not true that in all cases where (1) is true, (2) is true. A function of the heart's pumping blood is to circulate blood, but under certain abnormal conditions the blood's circulating can be caused by a pump of a certain kind. Hence, under these conditions the heart's pumping blood is not a necessary condition for the blood's circulating. Secondly, it is not true that whenever (2) is true, (1) is true. The heart's pumping blood is a necessary condition for the sound of the heartbeat, but it seems strange to say that a function of the heart's pumping blood is to produce the sounds of a heartbeat.

Perhaps the analysis of (1) can be improved. Perhaps (2) should be replaced by (3) as an analysis of (1):

> (3) X causes Y (under normal conditions), and Y is a necessary condition for the proper functioning of the organism.

The qualifying phrase "causes Y (under normal conditions)" would eliminate the problem of the pump replacing the heart. For in (3) it is not claimed that X is a necessary condition of Y. Although the heart's pumping blood is not a necessary condition for circulating the blood, the heart's pumping blood causes the blood to circulate under normal conditions. Furthermore, the phrase "is a necessary condition for the proper functioning of the organism" seems to eliminate the problem of the sound of the heartbeat. The sound of the heartbeat is not a necessary condition for the proper functioning of the organism. For it is presumably possible for the heart to operate silently and yet in another respect normally. (Suppose, for example, a heart "silencer" were made which, when attached to the heart, made the beat of the heart impossible to detect by normal sound apparatus. Or suppose that some new organism were found whose heart operated silently.) However, it is necessary to the proper functioning of the organism for the blood to circulate.

Nevertheless, despite the improvement in our analysis, (3) is still not completely adequate. For suppose there were a blockage of blood vessels to the heart, as happens with elderly people. In cases like this, the blood sometimes finds new paths to the heart. Now in these cases, the blockage of the blood vessels causes the blood to find new paths, and this is necessary to the proper functioning of the organism. However, it seems strange indeed to say that the function of the blockage of blood vessels to the heart is to develop new paths of circulation. This problem can be overcome by making the following qualifications:

> (4) X causes Y (under normal conditions) and X is not a malfunction of the organism and Y is a necessary condition for the proper functioning of the organism.

The qualifying phrase "is not a malfunction of the organism" would presumably eliminate cases like the blockage of the blood vessels to the heart, as well as similar cases, since blocked blood vessels are considered a malfunction of the organism.

Whether (4) is free from all counter-examples is difficult to say. Careful consideration of contexts in which functional statements are used may reveal cases in which the analysis is too broad or too narrow. Further qualifications and amendments may indeed have to be made.

In any case, there are problems with this analysis of a different sort. The analysis uses the word "cause." This term is notoriously unclear; moreover, as we have seen, as science advances, the notion of cause seems to be replaced by precise mathematical relations between variables. It has been suggested that a similar phenomenon occurs in biology with respect to functional statements:

> The tendency of modern biologists seems to be to restrict the use of function statements. Where strict [mathematical] functional relations can be found to hold among variables . . . function statements tend to not be used. Function statements are also used in textbooks on physiology. In fact, it is here that they are most widely used at present. These points regarding the usage of function statements suggest that the main importance of function statements is in the more practical, as contrasted with the more theoretical, concerns of biology.[21]

Waiving the problem of the obscurity of the notion of cause and the probability of the eventual replacement of functional statements by precise mathematical statements as biology develops, the question of whether functional statements have any explanatory value remains.

Hempel has argued that, in order for an activity such as that of the heart to be explained by citing a function of this activity, a D–N explanation of the following kind would have to hold:

L_1 For every organism of type O_1, if it is functioning properly, then its blood circulates.

L_2 The blood of an organism of type O_1 circulates if and only if its heart is pumping blood.

C_1 This organism is of type O_1.

C_2 This organism is functioning properly.

E This organism's heart is pumping blood.[22]

However, as Hempel points out, L_2 is not true since blood can circulate without the heart pumping blood, e.g., if the organism is equipped with an artificial "heart." Hempel argues that L_2 must be replaced by:

L_2' The blood of an organism of type O_1 circulates if and only if any one of a number of functionally equivalent activities occurs in it.

The heart's pumping blood would be one of these activities. Given L_2' and L_1 and C_1 and C_2, one could conclude

E' This organism has one or the other of these functionally equivalent activities occurring in it.

But this would not explain the heart's pumping blood in this organism, that is, it would not explain this *particular* activity.

On the other hand, Lehman has argued:

> There is a sense of "explain" in which one might speak of functional explanations. In a broad sense, any appropriate answer to a "why" question is an explanation and function statements may be used appropriately to answer why questions in certain contexts. Thus, if a medical student asks his teacher why certain animals have archinephric ducts, a correct answer, depending on the animal, is that the archinephric ducts serve the function of transporting sperm cells or urine (or both) out of the organism's body. In this case the teacher might be said to have answered with a functional explanation.[23]

Both Hempel and Lehman may be correct. Functional statements cannot be construed to provide a D–N explanation, and in *this* sense they do not provide an explanation. However, in another sense they do. The D–N model was meant to specify the logical and epistemological requirements for certain types of explanations found in science. As we have suggested above, it is an open question whether it specifies the logical and epistemological requirements for all explanations.

Whatever our final judgment about the explanatory value of functional statements in biology, it can hardly be denied that functional statements do help one begin to understand the phenomenon at issue. As we have seen, understanding a phenomenon scientifically involves seeing it in a nomological nexus. Knowing that a function of X is Y does provide one nomological connection. To be sure, other connections would have to be known to provide a more complete understanding of X.

Whatever the explanatory value of functional statements in biology, the question of their usefulness in biology education remains. In the next section we will consider this problem.

FUNCTION IN BIOLOGICAL EDUCATION

Jerome Bruner in discussing science education says:

> One of the principal organizing concepts in biology is the persistent question, "What function does this thing serve?" —a question premised on the assumption that everything one finds in an organism serves some function or it probably would not survive. Other general ideas are related to this question. The student who makes progress in biology learns to ask the question more and more subtly, to relate more and more things to it. At the next step he asks what function a particular structure or process serves in the light of what is required in the total functioning of an organism. Measuring and categorizing are carried out in the service of the general idea of function. Then beyond that he may organize this knowledge in terms of a still more comprehensive notion of function, turning to cellular structure or to phylogenetic comparison.[24]

I take it that one of Bruner's major theses in this passage is that biological research, as well as biological education, progresses by asking more detailed questions about function. As we have seen, however, in the science of biology, questions about function come in at the early stages of biological research and are virtually absent at the advanced stages. Thus the picture Bruner paints of biological research does not seem to be accurate. The same thing can be said of education in biology; as students develop biological sophistication, considerations of function will play a much less important role than they do at the outset.

However, this does not mean that approaching biological education through a consideration of function is not useful or worthwhile. There is good reason to suppose that one of the best ways to introduce students to biology is the way Bruner suggests, namely, by considering the functions of different parts of organisms. Indeed, perhaps part of the trouble with the Yellow and Blue Versions of B.S.C.S., and one of the advantages of the Green Version, has been the neglect of function and the inordinate stress on biochemistry and biophysics in the former and the wise use of function in the latter. Thus Ausubel has noted:

> The Green Version's policy of treating function first, and then showing how function is served by complementary structure, rather than adopting the opposite sequence of presentation, is very appropriate for the beginning student.

Functions can typically be remembered longer than structures. Thus, if function is stressed, more can be retained from which its complement can later be reconstructed, if forgotten, than if structure is emphasized. It is also easier to reconstruct forgotten structure from remembered function than vice versa.[25]

In any case, an introductory course in biology should be organized around those concepts and approaches with the most intuitive appeal and pedagogical value for beginning students, rather than around those typical of advanced research in the field. Pedagogical consideration here as in other places in education should be given more weight than technical accuracy and precision. The functional approach to biology, despite its limitations for advanced research, may well have great intuitive appeal and pedagogical value, and the concept of function may serve well as the major organizing concept in biological curricula. Philosophical analysis such as the preceding would then be of crucial importance to a biology curriculum developer. Such analysis would clarify the key concepts in beginning biological education.

•

CHAPTER THREE

Definition

•

Educators like Mr. Brown are interested in definition. He has become aware of the importance of operational definitions in science, as we have seen, through his reading of curriculum theorists like Bruner. Because of this he has begun to think about, instead of merely use, definitions. All science educators have used definitions in their work, even if in many cases they have not given them much thought.

Thus science teachers define terms to their class, science textbook writers define terms for their readers, science education researchers define terms in their research reports, and typically do not consider what exactly they are doing. For example, a typical high school physics teacher might define "conservation system" to his class. He would not consider what he was doing in any explicit way—for example, what sort of a definition he was giving, or the role of definitions in science education.

In this chapter we will consider definitions in science and science education. We will consider first some of the ambiguities in the notion of definition, noting different kinds of definition and their importance in science. Once we clarify some of the different types of definition, we will consider a recent argument that purports to show that scientific theories cannot be compared because different definitions are given for the same terms in different theories. After our general discussion of definition in science, we will go on to consider some of the uses of definition in science education. Toward the end of the chapter we will focus on one particular type of definition that has gained great popularity in physics: operational definition. We will analyze this type of definition and evaluate its usefulness in science. Finally we will consider the importance of operational definitions for science education.

DEFINITION AS AN EXPLICIT-STATEMENT OF SAMENESS OF MEANING

For our purposes, a definition will be considered a statement explicitly specifying that one word or expression (the word or expression to be defined) means the same thing as some other word or expression (the word or expression used to define the word or phrase to be defined).[1] For example:

(1) "Conservation system" means "a system of particles in which the forces acting on any particle of the system are forces which can be derived from a potential energy function."

(2) "x is magnetic" means "if a small piece of iron were placed near x, then the piece of iron would move toward x."

These are typical cases of definitions as we shall understand the term. For convenience we shall refer to the word or expression to be defined as the *definiendum* and the word or expression used to define the definiendum as the *definiens*. Thus in (1) above "conservation system" is the definiendum and "a system of particles in which forces acting on any particle of the system are forces which can be derived from a potential energy function" is the definiens; (1) states explicitly that the definiendum means the same thing as the definiens.

Definitions as they are conceived of here — explicit statements of sameness of meaning between definiendum and definiens — are the product of a human activity, what one might call defining activity. One of the most important uses of definitions is explaining the meaning of certain expressions to other human beings. However, stating definitions is not the only way of explaining the meaning of an expression to someone else. For example, a science teacher might explain the term "metal" to a child by pointing to cases in which the term correctly applies. We will call such a procedure *ostensive teaching of words*.[2] Again, a science teacher might explain the word "magnetic" to a student by using the word correctly in certain contexts. For example, he might say to his class,

"When something is magnetic and a small piece of iron is placed near it, the iron is pulled toward the magnetic object." We shall call such a procedure *teaching words contextually*.

In some cases whole expressions are defined. Instead of defining the term "magnetic" in isolation, or instead of teaching the meaning of a term contextually, a science teacher might define the expression "*x* is magnetic." An example of the resulting definition would be (2) above. Here the teacher has not defined "magnetic" but the phrase "*x* is magnetic." Such a statement—unlike the statement of the teacher in teaching the word "magnetic" contextually—is an explicit statement that certain words mean the same thing as other words. We shall use the term "definition" to refer to such a statement and refer to this explicit statement of the meaning of a phrase as a *contextual definition*. (We shall discuss one type of contextual definition of "magnetic" later when we consider operational definition.)

TYPES OF DEFINITIONS

What a person is intending to do in his defining activity of specifying a relation between the definiendum and definiens distinguishes different types of definitions.

First of all, when a person defines an expression he may be attempting to say something about the actual meaning of this expression. That is, he may be formulating a definition that is to be evaluated by how well it corresponds to some group's actual usage. In the case of scientific definitions, presumably this group would be the class of present-day scientists or perhaps some more restricted group, e.g., present-day physical scientists or present-day physical scientists who are careful and consistent in their use of language. Let us call definitions that purport to reflect actual usage *reportive definitions*. It is clear that such definitions are either correct or incorrect—or at least more or less correct or incorrect—and can be evaluated as such when it is known what usage the definition is supposed to reflect. In this case linguistic research may be very relevant.

However, someone may not intend his definition to reflect language habits at all. He may rather be stipulating a meaning. His definition then is tantamount to an announcement of how he proposes to use a particular expression in his lecture or paper, regardless of how it is normally used by scientists, if it is used at all. We shall call such definitions *stipulative definitions*. Stipulative definitions are neither correct nor incorrect, since they are proposals to use an expression in a certain way. Nevertheless, such definitions can be rationally evaluated. The formulation of a stipulative definition may be useful or misleading, helpful or confusing. Thus, although such definitions are neither correct nor incorrect, they can be evaluated as good or bad proposals. The relevance of stipulative definitions in science education and the criteria that should be used to evaluate such definitions in science education will be discussed later.

In scientific contexts it is sometimes difficult to tell without further information about the speaker or writer's intentions whether a definition is reportive or stipulative. Hence it is sometimes difficult to tell whether a definition proposed in scientific literature is incorrect because it does not reflect the way scientists use the definiendum or whether it is a proposal of how a scientist intends to use the definiendum.

There are some types of definition that do not fit easily into the category of reportive or stipulative definition. Sometimes the definer does not intend his statement to be a completely accurate description of ordinary usage, yet ordinary usage is not completely irrelevant to his purposes. He intends his statement to be in some way an improvement over the ordinary meaning of a term.

For example, a person may intend that his definition make a vague and imprecise term less vague and imprecise. Let us call a definition that purports to reduce the vagueness of the ordinary meaning of a term an *explication.*[3] Such a definition would be judged in terms of how well that part of the ordinary meaning of a term that is not vague is retained and how well that part of the ordinary meaning of a term that is vague is reduced.

Not all improvements in the meaning of terms for scientific purposes are improvements in precision, however. The reportive definition of a term may be modified for scientific purposes for various reasons, lack of precision being only one. Thus explication is a special case of improving the meaning of a term for scientific purposes. The sort of definition that improves upon reportive definitions for scientific purposes will here be called *rational reconstruction.* Explications are one type of rational reconstruction.

A rational reconstruction is to be evaluated not solely in terms of standard usage, but also in terms of its ability to improve upon such usage for scientific purposes. Depending on the inadequacies of standard usage for scientific purposes, a rational reconstruction will depart more or less from a reportive definition. It is sometimes difficult to tell in scientific contexts whether a proposed definition is intended to be a reportive definition or a rational reconstruction. However, sometimes it is clear that a rational reconstruction is intended. There are many examples of this in the history of science. When Newton defined "force" as "the product of mass times acceleration," it is unlikely that he thought he was reporting the standard scientific meaning of "force" or purely stipulating some arbitrary meaning of "force." The most plausible interpretation is that he thought he was perfecting the standard meaning of "force" in such a way that it was more precise and fit in with developments of science in his own time—most notably the development of his own theory.[4]

ANALYTIC AND EXTENSIONAL DEFINITIONS

Whether a definition is intended to be reportive or stipulative or a rational reconstruction, the question remains of what sort of relation holds between the definiendum and the definiens. There are two possibilities—that the definiens is meant to specify the sense or connotation of the definiendum, or that the definiens is meant to specify the extension or reference of the definiendum. If the sense meaning is specified, we have an *analytic definition*. On the other hand, if the extension of the definiendum is specified, we have an *extensional definition*.[5]

Since the categories of analytic definition and extensional definition cut across the distinction between reportive definition, stipulative definition, and rational reconstruction, we have six different types of definition: analytic reportive definition, analytic stipulative definition, analytic rational reconstruction, extensional rational reconstruction, extensional reportive definition, and extensional stipulative definition.

Let us consider analytic and extensional definitions in more detail.

In an *analytic definition*, the definiens is used either to specify the sense or connotation of the definiendum among some groups of language users (reportive), or to specify the sense or connotation of the definiendum the speaker or writer intends the definiendum to have (stipulative).

The sense or connotation of a term is usually considered to be the properties or characteristics that are semantically relevant for determining whether the term should apply. To say that property P is semantically relevant for determining whether term T applies is to say that P's presence or absence counts in and of itself to some extent in determining whether T applies or does not apply.[6] Thus, having a sour taste is semantically relevant for determining whether the term "lemon" applies; the presence or absence of the property counts in and of itself to some extent in determining whether the term "lemon" applies. Hence, having a sour taste is part of the sense or connotation of "lemon." On the other hand, the presence or absence of the property of growing in Florida would not count in and of itself in determining whether the term "lemon" applies or does not apply. The presence or absence of the property of growing in Florida would count in determining whether or not the term "lemon" applied only because of the connection of this property to other properties. Hence, the property of growing in Florida is not semantically relevant and is not part of the sense or connotation of "lemon."

Sometimes the properties that are semantically relevant are logically necessary conditions for the application of a term. For example, the property of being a man is a logically necessary condition for the application of the term "brother," since if something did not have this property the term "brother" could not be applied. At other times the properties that are semantically relevant are logically sufficient conditions for the application of the term. For example, the property of being a figure enclosed by three straight lines is logically sufficient for the application of the term "triangle," since if anything has this property then the term "triangle" can be applied.

Sometimes, however, the properties that are semantically relevant are neither logically necessary nor logically sufficient conditions for the correct application of a term. The property of being sour, for example, is not a logically necessary condition for the application of the term "lemon," since it is logically possible that there could be sweet lemons. That is, English speakers would apply the term "lemon" to a fruit even though it were sweet, given that the fruit had many other semantically relevant properties, e.g., yellow color, waxy texture.

To say, then, that the definiendum has the same sense or connotation as the definiens is to say that the definiens specifies those properties that constitute the sense or connotation of the definiendum, i.e., those properties that are semantically relevant for the application of the definiendum. These properties sometimes, but not always, are logically necessary or logically sufficient conditions for the application of the definiendum.

Let us examine how these considerations operate in the context of scientific definitions. A scientist might put forth the following definition:

> "Rigid body" has the same meaning as "a body in which the distance between any two points remains constant."

Suppose the scientist intends the definition to reflect the meaning of "rigid body" among physicists. Then it is a reportive definition. Moreover, he intends the definition to reflect the sense of "rigid body" among physicists. (Hence the expression "has the same meaning as" is intended to refer to the sense or connotation of the terms at issue.) Then the definition is an analytic definition. The definiens purports to specify those properties that are semantically relevant to the application of the definiendum "rigid body," i.e., the property of being a body in which the distance between any two points remains constant. The definition seems to be correct; the definiens does specify the sense or connotation of the definiendum. In this case the property seems to be both a logically necessary and a logically sufficient condition for the application of the term "rigid body" as that term is used among physicists.

However, consider the following definition:

> "Copper" means the same as "a metallic element with atomic number 29, that is reddish brown, a good conductor of heat and electricity, with a melting point of 1083°C."

Let us suppose that this definition is also intended to specify the sense or connotation of the term "copper" among physical scientists. It is quite likely that this definition is correct; the definiens does specify the sense of the term "copper" as that term is used among physical scientists.

However, the properties mentioned in the definiens are neither logically necessary nor logically sufficient conditions for the application of the term "copper" by physical scientists. For example, given drastic changes in atomic theory, scientists would probably still call a piece of metal "copper," provided many other of the semantically relevant properties were present. Hence, having atomic number 29 is not a logically necessary condition for the application of the term "copper." Moreover, it would probably be the case that a metal which had atomic weight 29, but which was not a good conductor of electricity and was purple, would not be called "copper." Hence, having atomic weight 29 is not a logically sufficient condition for the application of the term "copper." One might ask whether the properties specified by the definiens would constitute jointly a logically sufficient condition for the application of the term "copper." Whether it would is at least debatable. It is unclear if a scientist would call a piece of metal "copper" that had atomic weight 29, was reddish brown and a good conductor of electricity, and so on, if at the same time its size and weight varied in unpredictable ways and it gave off strange musical sounds.

The above discussion can perhaps be summarized in this way. For some scientific terms it is possible in a reportive definition to specify logically necessary and sufficient conditions for the application of the term. In other cases this is impossible. The best that can be done in some cases is to specify those properties that are semantically relevant for the application of the term. For stipulative definitions there is no problem. It is always possible to stipulate a sense or connotation of a term that is both logically necessary and sufficient for the application of the term.

An *analytic rational reconstruction* of a term would be a definition that improves upon the sense or connotation of the term for scientific purposes. This improvement could be accomplished in several ways. First, it might not be clear in the standard use of a term whether a property was semantically relevant, that is, whether the presence of *A* counted to some extent in and of itself in determining if a term applied. In an analytic rational reconstruction this problem could be settled, for example, by specifying only those properties which were clearly semantically relevant, or by making a decision to include those properties which were unclear either in the class of semantically relevant properties or in the class of semantically irrelevant properties—a decision which would be made on the basis of the usefulness of the resulting rational reconstruction to scientific theory and practice.

Secondly, as we have seen, certain properties are neither logically necessary nor logically sufficient conditions for the application of a term, although they are semantically relevant. Because of this, it may be difficult at times to decide when a term actually applies, when in a given case some of the semantically relevant properties are absent and some of the semantically relevant properties are present. Consider, for example, a piece of metal that does not have atomic number 29 and does not conduct heat well, but looks exactly like copper, conducts electricity well, and melts at 1083°C. An analytic rational reconstruction of a term might make a certain semantically relevant property a logically necessary or sufficient condition for the application of a term, even though it was not a logically necessary or sufficient condition before. For example, although atomic number 29 is not a logically necessary condition for the application of the term "copper," it is semantically relevant for such application, perhaps more semantically relevant than any other property. In a rational reconstruction of "copper," having atomic number 29 might be made a logically necessary condition for the application of the term "copper." Given this reconstruction, the above-mentioned piece of metal would not be copper. Such a change might be justified on several grounds. It might be argued that, in modern chemistry, atomic number has become increasingly important and such a change in definition is simply projecting a current tendency, or that it is simpler and easier to construe copper in this way, or that such a construal would eliminate areas of vagueness in the extension of the term "copper."

Analytic definitions, as we have seen, specify the sense or connotation of the definiendum. We have attempted to clarify the sense or connotation of a term in terms of semantically relevant properties. There is a rather close connection between analytic definitions as we have construed them and the traditional idea of analytic statements. Traditionally a statement was considered to be analytic if it was true by virtue of the sense of the terms used in the statement, and not by virtue of any nonlinguistic facts in the world. Thus, "Triangles have three sides" is considered to be analytic since it is true simply by virtue of the sense of the terms used in the sentence. On the other hand, statements which are true by virtue of some nonlinguistic facts in the world are called *synthetic statements*. An example of this type of statement is "Lemons grow in Florida." This is true not by virtue of the sense of the terms in the sentence, but because of certain nonlinguistic facts.

The relation between analytic definitions and analytic statements should then be obvious. A sentence like "All triangles have three sides" is used to express an analytic statement if and only if it is used in such a way that having three sides is part of the sense of "triangle," i.e., that "having three sides" is part of the definiens. Similarly, "Lemons grow in Florida" is used to express a synthetic statement if and only if it is used in such a way that "growing in Florida" is not part of the sense of "lemon," i.e., that "growing in Florida" is not part of the definiens.

However, analytic definitions inherit the problems of analytic statements. The notion of analyticity has come under strong attack in recent years.[7] Some prominent logicians and semanticists have argued that the notion is intrinsically unclear. In our account it is not difficult to see what some of the lack of clarity might be. We have suggested that a semantically relevant property of a term is a property the presence or absence of which counts in and of itself toward determining whether the term applies. However, although one has some intuitive grasp of what "in and of itself" means, it is difficult to make this idea clear. A less problematic type of definition would not use the sense or connotation of a term at all in specifying the meaning of the term.

In contrast to analytic definitions, there are definitions in which the definiens does not specify the sense or connotation of the definiendum, but rather the extension or reference of the definiendum. These are called *extensional definitions*. Now the extension of a term is simply the class of things the term applies to, by virtue of either standard usage (reportive) or arbitrary fiat (stipulative). Consider the reportive definition of "man" as "featherless biped." The terms "man" and "featherless biped," when used by English speakers, apply to the same class of things. Thus these two terms have the same extension. However, "featherless biped" and "man" do not have the same sense or connotation. The presence or absence of the property of being a featherless biped presumably does not in and of itself count toward determining whether the term "man" applies or does not apply; the property of being a featherless biped is not semantically relevant for the application of the term "man." If the presence or absence of such a property counts at all, it is by virtue of its connection with other properties, e.g., the property of being rational. The property of being rational does seem to be semantically relevant for the application of the term "man."

It is sometimes difficult in scientific contexts to tell whether a reportive definition is an incorrect analytic definition or a correct extensional definition. For example, a scientist might say that "red" has the same meaning as "light ray of wavelength 6500 to 7000 A." Although it is doubtful that the definiens specifies the sense of "red" even among scientists — it certainly does not among laymen — the definiens does have the same extension as "red." Thus, although the scientist's definition is probably an incorrect analytic reportive definition of "red," it is a perfectly correct extensional definition. Hence, unless we are clear on what the scientist was attempting to do in his definition, it is difficult to evaluate it.

Extensional definitions in scientific contexts are very important. First, although it may be impossible in certain cases to give a reportive analytic definition of some scientific term, it may not be impossible to give an extensional definition. Color words like "red" and "blue" are often said to be indefinable. What is usually meant by this is that it is impossible to construct a definiens for the word "red" or "blue" that specifies the sense or connotation of "red" or "blue" as these words are used among English speakers, except for the completely uninformative, "red" means the same as "red." (There would, of course, be no problem in giving a stipulative definition of "red" or "blue.") Whether this claim is true or not, it does seem to be possible to give an extensional definition of "red" in terms of wavelengths that is quite informative.

Moreover, deduction, contradiction, and other important logical properties of scientific propositions and theories turn on extensional relations between terms, and not on sense or connotation.[8] As we shall see shortly, this point is important in meeting a recent attack on the objectivity of science. In this sense at least, extensional definitions are much more important scientifically than analytic definitions.

Thirdly, as we have seen, the sense or connotation of a term has seemed to many logicians and semanticists to be much less clear and much more problematic than the extension of a term. Accordingly, analytic definitions are much more problematic than extensional definitions, which use the relatively clear notion of extension. Hence extensional definition has still another advantage in scientific contexts.

An *extensional rational reconstruction* of a term is a definition which improves on the extension of the term for scientific purposes. This improvement can be accomplished in several ways, depending on the term and/or the purposes of the reconstruction. First, the extension of a term might be vague, that is, there might be a wide range of cases in which it is impossible to tell whether the term applies or not. A rational reconstruction of the term might specify the extension of a term in such a way that this area of vagueness is eliminated or at least decreased.

For example, an economist might give the following definition of the term "wealthy American": "a. U.S. resident who makes $25,000 or more a year." This rational reconstruction would make the extension of the term "wealthy American" much more precise than it now is among American users of the English language. For, as the term is used, there is a large area of vagueness, a large number of cases in which it is impossible to say whether or not the term "wealthy American " applies. This rational reconstruction would tend to eliminate the area of uncertainty.

Whether this is an adequate rational reconstruction is another question. Although it may eliminate the vagueness, it may also eliminate clear cases of wealthy Americans that the economist does not want to eliminate. For example, a person who made only $20,000 a year but owned $100,000 in AT&T Common Stock would normally be considered wealthy. Whether such a case would show the inadequacy of the economist's rational reconstruction would depend a great deal on what the economist wanted to do with his rational reconstruction, what function it had in economic theory and practice.

Secondly, the extension of a term might not be vague and yet there might be theoretical reasons for modifying it. Suppose the term "reasoning" is applied by English speakers only to certain activities of human beings and never to the activities of animals. However, suppose that discoveries in animal psychology indicate that the properties constituting the sense of "reasoning" are also found in certain animals, e.g., dolphins. There might be good reason to broaden the extension of the term to include certain activities of some animals. In this case a rational reconstruction is produced not to cure vagueness, but to cure a narrower range of application than is justified by scientific findings.

Closely related to an extensional rational reconstruction, as we conceive of it here, is an *extensional isomorphic definition.*[9] As we have seen, an extensional definition specifies that a definiens has the same extension as the definiendum. However, substituting the definiens for the definiendum does not always preserve the truth value of the sentence in which it is substituted. For example, consider the sentence:

> (1) John believed that he saw the morning star high in the sky.

The term "morning star" has the same extension as the term "evening star." Consider the sentence:

(2) John believed that he saw the evening star high in the sky.

(1) may be true and yet (2) may be false. John might believe the one without believing the other, since he does not believe that the extension of "morning star" is the same as the extension of "evening star," i.e., he does not believe that the morning star is the evening star. Let us call contexts like the above, in which substitution of terms with the same extension does not preserve truth value, *nonextensional contexts*. Then in all extensional contexts, substituting terms with the same extension is truth preserving.

It follows that substituting the definiens for the definiendum from a correct reportive extensional definition is truth preserving in all extensional contexts. However, consider a definition in which (with respect to the ordinary meaning of the definiendum) substituting the definiens for the definiendum is truth preserving only in *some* extensional contexts. Such a definition would, of course, be an incorrect reportive extensional definition. Nevertheless, such a definition might be extremely useful for science. For the contexts in which substitution is truth preserving might be the only contexts one is interested in, given one's theoretical purposes. For example, assume "wealthy American" means "a U.S. resident who makes $25,000 or more a year." Substituting the definiens for the definiendum may not be truth preserving in all extensional contexts, if we take the two expressions in their usual meaning. However, substituting may be truth preserving in all the extensional contexts an economist might be interested in. A statement of an extensional isomorphic definition, then, purports to specify a relation between the definiendum and definiens such that substituting the definiens for the definiendum preserves truth in some contexts, namely, those contexts in which one has theoretical interest.

DEFINITIONS, MEANING CHANGE, AND SCIENTIFIC OBJECTIVITY

In view of our analysis of different types of definitions, we should be able to evaluate a recent argument against the possibility of scientific objectivity. We saw in Chapter One that two rival theories in a given domain are evaluated by deducing contradictory empirical consequences from these two theories (plus auxiliary hypotheses) and then seeing which of the conflicting consequences corresponds with the facts. This account, some people argue, presupposes that the meanings of the terms in the two theories are the same, for if they are not the same, the two consequences could not be in conflict.[10] These people maintain that, as a matter of fact, the meanings of terms do change from one theory to another. Hence, they conclude that two theories in a given domain cannot be evaluated by comparing their deduced empirical consequences. Since this has been considered the standard means of objectively evaluating two theories in a domain, a change of the meaning of a scientific term from one theory to another is said to affect scientific objectivity seriously.

Consider, for example, two rival atomic theories, T_1 and T_2. Suppose that T_1 (plus auxiliary hypotheses) entails:

(1) If anything is an atom, then it has property P.

And T_2 (plus auxiliary hypotheses) entails:

(2) Something is an atom and does not have property P.

It certainly seems that (1) and (2) are logically incompatible, hence that both could not be true. Thus, if the auxiliary hypotheses are assumed to be true, T_1 and T_2 are in conflict and either T_1 or T_2 is eliminated, depending on whether (1) or (2) is false.

However, suppose that the term "atom" in (1) does not have the same meaning as the term "atom" in (2), and suppose that "property P" in (1) does not have the same meaning as "property P" in (2). Then, it has been argued, (1) and (2) would not be logically incompatible. Furthermore, suppose that all the terms that seem to be the same and are found in both theories mean something different in each theory. Then T_1 and T_2 could not be in conflict and could not be evaluated against one another in the standard way.

This argument, if it were sound, would be a powerful one against the possibility of an objective evaluation of rival theories in a domain. However, the argument is not sound. We can approach the argument by means of definitions. First let us suppose that the senses or connotations of all the terms in T_1 that are also found in T_2 are different. Let us suppose that all the analytic reportive definitions of terms in T_1 are different from the analytic reportive definitions of those terms found in T_2. For example, the analytic reportive definition of "atom" in T_1 is different from the analytic reportive definition of "atom" in T_2. Thus, in one sense of "meaning," namely, the sense or connotation of a term, all terms in T_1 have changed meaning in T_2.

However, difference in sense or connotation is perfectly compatible with sameness of extension. Thus differences in sense meaning are perfectly compatible with sameness of extensional meaning. However, sameness of extension is all that is needed for two theories to be logically incompatible.[11]

To return to our example: Even if "atom" in (1) has a different analytic reportive definition from "atom" in (2), and "property P" has a different analytic reportive definition in (1) from "property P" in (2), it well might be that "atom" in (1) has the same extensional reportive definition as "atom" in (2), and "property P" in (1) has the same extensional reportive definition as "property P" in (2). If this were true, (1) and (2) would be logically incompatible. This is seen clearly when we realize that to say (1) and (2) are logically incompatible is merely to say (1) and (2) cannot both be true. And they cannot both be true if the extension of "atom" and "property P" does not change from T_1 to T_2.

We can conclude that change in the sense or connotation of scientific terms in rival theories is perfectly compatible with comparing two rival theories in terms of their empirical consequences. However, what if the extensions of the terms in T_1 and T_2 change as well? Suppose, for example, that the extensional reportive definitions of "atom" and "property P" vary from (1) to (2). Could T_1 and T_2 still be in conflict?

They could so long as the extension of the terms stood in a certain relation.[12] For purposes of clarity, let "atom$_1$," and "property P_1," be the terms that appear in (1) and "atom$_2$," and "property P_2," be the terms that appear in (2). So construed, (1) and (2) become:

(1') If anything is an atom$_1$, then it has property P_1.
(2') Something is an atom$_2$ and does not have property P_2.

Let us suppose that the following auxiliary hypotheses are true:

(3) If anything is an atom$_2$, then it is an atom$_1$.
(4) If anything has property P_1, then it has property P_2.

(3) and (4) say, in effect, that the extension of "atom$_2$," is included in the extension of "atom$_1$," and that the extension of "property P_1," is included in the extension of "property P_2." (This is perfectly compatible with the extension of "atom$_1$," and "atom$_2$," not being identical and the extension of "property P_1," and "property P_2," not being identical.) However, (1'), (2'), (3), and (4) are logically incompatible. Hence T_1 and T_2 are in conflict relative to auxiliary hypotheses (3) and (4). We can conclude, therefore, that even if the extensions of terms change from one theory to another, i.e., even if the reportive extensional definitions of terms are modified from one theory to another in at least some cases, the theories can still be compared logically and evaluated objectively. Meaning change, therefore, of either connotation or extension is not necessarily a hindrance to the objective comparison and evaluation of two theories.

DEFINITION AND SCIENCE EDUCATION

The above discussion of definition has many applications for science educators. We will consider here only three areas in which definitions can play an important role in science education. We will also consider some applications of our discussion for science textbook writers, researchers in science education, and science teachers.

The term "definition" often appears in science textbooks. Yet it is usually not clear what sort of definition is being discussed. One suspects that this lack of clarity is the result of the textbook writer's vagueness about the notion of definition. Furthermore, not only is it unclear what the science textbook writer is getting at in his discussion of definition, but, insofar as one understands it, the discussion seems to be based on an incorrect concept of definition.

Consider a passage from B.C.S.C.'s *Biological Science.*

> When you think of "life," what first comes to mind? Is it movement, a rose, breathing, animals, a beating heart, or something else you associate with things that are alive? Think about this question and then define "life" as carefully as you can.
>
> Now let us test your definition. Will it apply to all animals and all plants that you know? Will it apply to tiny creatures such as *Plasmodium* mentioned in Chapter 1? . . . Will your definition distinguish between the pea, the egg, and the potato when alive and after being cooked? . . . The chances are that you did not succeed in making a definition of life that would work in *every* case. No one has ever done so.[13]

How is one to understand the term "definition" in this passage? One supposes that what is being asked for under the label of a definition is a reportive definition. However, even if this is granted, it is unclear whose usage the definition is supposed to reflect. Is it supposed to reflect the students'? English-speaking laymen's? Biological scientists'? Furthermore, are the authors of *Biological Science* talking about an analytic reportive definition or an extensional reportive definition? Their discussion suggests that an analytic reportive definition is at issue, but this is not clear. In any case, the authors conclude that a definition of life that "works in *every* case" has never been given. However, as we have seen, although in some cases an analytic reportive definition is impossible, an extensional reportive definition can be given. Thus, even if the authors are correct that an analytic reportive definition of life has never been given, an extensional reportive definition might be available.

Let us suppose that the authors have in mind an analytic reportive definition that reflects the usage of the biologist. Then
they seem to have too narrow a view of what such a definition
consists of. They seem to suppose that the definition must specify
some property that is a logically necessary and sufficient condition
for the application of the term "life." In discussing whether movement is the chief characteristic of life they say:

> We would discover that the "ability to move" is not a very
> satisfactory definition of life. Some things that are alive do
> not move. Some things that move are not alive. If we are to
> define life adequately, all things that are alive must meet
> our definition and no things that are not alive can be in-
> cluded.

Although this may be true, "the ability to move" may still be
in many contexts a semantically relevant property of "life," and
thus *part* of the connotation of the term "life." As we have seen,
not all defining properties of a term, when analytic definitions are
at issue, need be logically necessary or sufficient conditions for the
application of the term. Small wonder, with this strict criterion of
what could constitute a definition of life, that the authors claim
there is no definition of life that "would work in every case."

The authors then go on to say, "We can work towards a definition of life that is adequate for nearly all purposes by asking
some more questions." At this point the authors appear to have
moved into a different type of definition—a definition that is
judged not in terms of whether it reflects all cases of ordinary use,
but in terms of whether it is "adequate for nearly all purposes."
Here they seem to be talking about a rational reconstruction of
the term "life"—a definition of life that is an improvement on the
ordinary meaning for certain scientific purposes.

The discussion of the definition of life in this textbook could
have been greatly improved if the authors had clarified the various definitions at issue. Such clarity might have suggested a different approach to the material. Instead of immediately asking
that the student give a definition of "life," it might have been useful to have a short discussion on definition in biological science:
different kinds of definition, criteria for evaluating them, and so
on. Students would then be asked to define "life"; in particular,
they would be asked to give an analytic reportive definition of
"life" in their own terms. The problems with such a definition
could then be shown and the following chapter could attempt to
give an extensional rational reconstruction of "life" that would
meet the problem of the definition given by the students.

RESEARCHERS IN SCIENCE EDUCATION

As we have seen, learning definitions is not the only way of understanding the meaning of a term in science. One might teach the meaning of a term ostensively as well as contextually. The problem remains of finding the best way to get students to understand the meaning of scientific terms. Different hypotheses suggest themselves and these hypotheses could be tested by researchers in science education. One hypothesis is:

H_1 Better understanding of the meaning of scientific terms in young and immature science students is achieved by ostensive teaching of the meaning of the terms, rather than by definitions of the meaning of the terms.

Another hypothesis is:

H_2 Better understanding of the meaning of scientific terms is achieved in smaller classes by the contextual teaching of these terms, rather than by definition of the meaning of the terms.

Perhaps it is more effective to have the teacher define scientific terms for the student by giving them reportive definitions than to have the students discover these definitions, e.g., by reading the text and constructing a definition from the reading. One hypothesis would be:

H_3 Better understanding of the meaning of scientific terms is achieved by students discovering the definition of a scientific term, rather than by the teacher defining the term for them.

It should be noted that, even if H_3 were well supported by research findings, it would not necessarily follow that teachers should use the discovery method in teaching students definitions. Such a method might take a long time or be very expensive. The increased understanding might be too small to justify the increased time and expense. But even here educational research would be useful. For research could be done to determine the length of time it takes a group of students to discover a definition in science, e.g., by reading a scientific textbook and constructing one from the examples in the text, in contrast to the length of time it takes teachers to define a scientific term for a student.

All of the above hypotheses, H_1, H_2, and H_3, use the notion of understanding the meaning of a scientific term. Educational researchers might help develop adequate testable criteria for determining whether a student understands the meaning of a scientific term. Scientific educators have developed a test for whether students understand science; another test for whether students understand scientific terms should also be developed. Indeed, developing such a test would be necessary before the hypotheses suggested above could be tested. Some of the above discussion of a definition would be relevant for constructing such a test, for one way of looking at the problem of developing adequate criteria for determining when a student understands a term would be in terms of developing an adequate definition of "understanding the meaning of a scientific term." Indeed, the problem seems to come down to providing an adequate rational reconstruction of the phrase "understanding the meaning of a term" for the purposes of testing. A rational reconstruction of "understanding the meaning of a scientific term" would then be needed to test H_1, H_2, and H_3.

SCIENCE TEACHERS

Science teachers may depart from the accepted scientific meaning of scientific terms in order to communicate to their students. Not only should science teachers simplify the meaning of scientific terms for young children in order not to lose their attention, but they may be justified in giving stipulative definitions of scientific terms that depart radically from accepted usage. It may be good teaching and not sloppiness or intellectual negligence to say to a class of first-graders, "Children, for our purpose 'ion' means 'a tiny speck of stuff that pushes or pulls on other tiny specks of stuff.' "

This has not always been understood and it has sometimes been suggested that science teachers have the responsibility of conveying the precise meaning of scientific terms in actual scientific practice to their students. Thus one writer has said:

> The definitions of the fundamental concepts of the physical sciences are the lifeblood of those sciences. We must understand them ourselves with utmost clarity and must pass on to our students a similarly clear and (how I hate the word) unambiguous understanding of their meaning.[14]

Nevertheless, the same writer maintains that students should not memorize definitions if they can express the same concepts in their own words. But young science students can often only express these concepts in their own words in a way that distorts or at least oversimplifies the actual meaning. And it would certainly seem to be bad pedagogy in this case to require the students to memorize the correct definition, for it would not be comprehensible to them. It is far better that the students either receive or discover definitions that they can understand, even if these definitions are far from accurate. As the student matures intellectually, more scientifically acceptable meanings should, of course, be conveyed.

Stipulative definitions have other functions in science teaching. Not only can they be used to stipulate a new meaning of well-known scientific terms for the pedagogical purposes at hand, but they can be used to introduce a new term or to give a well-known nonscientific term a special meaning for the purposes at hand. The thoughtful science teacher, for example, may christen a particular phenomenon "discovered" in the laboratory or classroom with the name of the students who discovered it. Suppose young Dickie Robin discovers that when a cold jar of clear water is placed on top of a jar of warm water colored with ink so that the openings of two jars are fitted together, the colored water starts to go into the jar of clear water. The teacher may take advantage of this by a quick stipulative definition and a question: "By the 'Robin effect' let us understand 'the movement of the dark warm liquid from the bottom jar to the top jar.' Who can tell me what the cause of the Robin effect is?"

A funny but memorable new term may be used to refer to certain phenomena. "The Jobberwackey effect" and "the Teetledumb movement" may be amusing and easy-to-remember expressions for referring to certain physical phenomena that might otherwise be colorless and insignificant to the students. The science teacher need not worry that these terms will be carried over into adult life; they will soon be forgotten as the child matures and their usefulness is past.

As the science student matures, the science teacher can not only convey the meaning of scientific terms via definitions to the student, but also teach the student how to define scientific terms and evaluate definitions. As a student learns, for example, to write scientific research papers and laboratory reports, he will need to define certain terms. As a critical reader of the research reports of others, he will need to evaluate the definitions given in these reports. Thus, part of the science student's training should be aimed at developing skill in defining, and many of the considerations involved in defining outlined above will be relevant.

For example, the budding young scientist should be able to state correctly what certain scientific terms mean among scientists, i.e., he should be able to give correct reportive definitions. But he should also be able to modify these definitions according to scientific need (rational reconstruction), use them in special senses for the purposes at hand, and introduce new terms when it is useful (stipulative definition). He should be able to state his definitions in such a way that his reader or listener is able to determine whether he is giving a reportive definition, stipulative definition, or rational reconstruction, an analytic or extensional definition, and so on.

Usually there is no special emphasis on developing the skill of defining in the training of scientists. Although a science student may become very proficient in setting up experiments, testing hypotheses, and so on, through special and systematic training, any skill he may develop in defining is picked up piecemeal and unsystematically. As a result, scientists are usually better experimenters than definers. A valuable experiment is often presented in a research paper that is confused and muddled, either because of unclear and faulty definitions of the key terms, or because of no definitions at all. As in any activity, skill is best achieved through systematic practice under the guiding hand of a teacher who has already mastered the skill. It is to be hoped that such systematic training will become part of science education.

OPERATIONAL DEFINITIONS

P. W. Bridgman is generally considered the founder of a methodological position known as operationism, which maintains that all scientific terms should be defined in terms of physical operations and the results of the operations used in deciding whether these terms correctly apply. Bridgman, however, always emphasized that he was merely making explicit a methodological point of view that was implicit in physics all along, and what Bridgman said may be true. Physicists—especially experimental physicists—have emphasized the overt physical manipulations involved in testing hypotheses and have sometimes tended to associate these manipulations and their results with their understanding of the hypotheses. It was a natural step, although perhaps a mistaken one, for Bridgman to identify the meaning of the terms in physics with physical manipulations and their results.

The definition of a scientific term in terms of physical manipulations and the results of these manipulations can be illustrated by the following contextual definitions. Consider the definiendum:

(1) x is magnetic.

Then the definiens might be:

(2) If a small piece of iron were placed near x, then the piece of iron would move toward x.

Again consider the definiendum:

(2) x is acid.

Then the definiens might be:

(3) If a piece of blue litmus paper were inserted into x, then the paper would turn red.

In general, the definiens in an operational definition has the following form:

If operation O were performed, then result R would occur.

Bridgman's position became more complicated and confusing in his later writing.[15] He suggested that nonphysical operations, variously called "verbal," "mental," and "paper and pencil" operations, could figure in operational definitions. It is quite unclear how these nonphysical operations were to define scientific terms, however, or how they fit into the operationist program. In any case, the dominant theme of Bridgman's early work, *The Logic of Modern Physics*, was the necessity of defining scientific terms in terms of physical operations, and it is this theme that is usually associated with operationism.[16]

The methodological program presented in Bridgman's book can be formulated by the following methodological rule:

MR_1 Define all terms in physics operationally.

If carried out in practice, this program would mean that the definiens for all physical terms could be specified in terms of physical operations and the results of these operations.

Bridgman sometimes tried to justify MR_1 by saying that, unless an operational definition of a scientific term were possible, the term was meaningless. However, it should be emphasized that one might accept MR_1 without accepting Bridgman's reason for MR_1. For example, one might argue that, although a scientific term like "magnetic" would not be meaningless unless an operational definition were possible, it would not be useful for science.

This brings us to the problem of what operational definitions are. It is unlikely that Bridgman or his followers conceived of operational definitions as stipulative definitions. First, Bridgman seemed to think that certain terms were incapable of operational definition and, hence, were meaningless. However, it is always possible to stipulate a definition for a term in terms of operations. Secondly, opponents of operationism have argued that it is impossible for the operational program to be carried out. However, if stipulative definitions were involved, there would be no difficulty in giving all physical terms operational definitions.

It is much more likely that Bridgman conceived of operational definitions as reportive definitions or as rational reconstructions. Let us consider these alternatives in turn.

Are operational definitions of physical terms reportive definitions? As we have seen, the correctness of reportive definitions is relative to some group of language users. If operational definitions are reportive definitions, presumably the relevant group of language users would be physicists. Supposing this to be so, we still have the problem of determining the sort of reportive definition at issue, analytic or extensional.

Let us consider analytic definitions first. Is the sense or connotation of (1) for physicists specified by (2) above? It is unlikely that it is if we are talking about the *entire* sense or connotation of (1). To be sure, attracting iron is a semantically relevant property for determining whether a thing is magnetic. But this surely is not the entire connotation of "magnetic" as this term is used by physicists, since other properties seem to be semantically relevant to the application of the term.

Moreover, in the case of some physical terms it is unlikely that even part of the sense or connotation can be specified by citing operations. Consider the term "diatomic molecule" as it is used in physics. Its sense or connotation does not seem in any obvious way to involve physical operations and the results of these. This does not mean, of course, that physical operations and the results of such operations are not relevant in determining whether a term like "diatomic atom" applies. The results of certain tests might, in conjunction with other assumptions, enable one to say that something was a diatomic molecule. What is at issue is whether the results of these tests are relevant *in themselves* (semantic relevance) or relevant because they are associated with the presence of other properties that are relevant in themselves (nonsemantic relevance). What is being questioned here is whether physical operations and their results are part of the sense of all terms in science, not whether they indicate the presence of other properties that are part of the sense.

The implausibility of construing operational definitions as analytic reportive definitions is brought out still further by another aspect of Bridgman's position. Bridgman maintained that one term is sometimes defined by two different operations and their results. Consider the definiendum:

(1′) *x* is magnetic.

Then the definiens might be:

(2') If x were moved through a closed wire loop, then an electric current would flow in the loop.

Bridgman would maintain that the meaning of "magnetic" in (1) is different from the meaning in (1'), and that, ideally, different words should be used to mark the ambiguity of "x is magnetic." Thus one might speak of "magnetic$_1$," or "iron-attraction magnetic" and "magnetic$_2$," or "wire-loop magnetic." If one is concerned with the sense or connotation of "magnetic" in actual scientific discourse, however, it is implausible to suppose that "magnetic" in (1) and (1') has different senses or connotations. The term "magnetic" does not seem to be ambiguous in these different contexts.

Thus we may conclude that Bridgman's operationist program, if interpreted in terms of analytic reportive definitions, is impossible; it is impossible to follow MR_1. What about extensional reportive definitions? Could operational definitions be construed in these terms? Could, for example, (2) above specify the extension of (1)? Strictly speaking it could not, for the definiens in an extensional definition is supposed to specify a condition that is necessary and sufficient for the application of the definiendum, although it need not be logically necessary or sufficient. But this is surely not true for (2); it is not always the case that if x is magnetic, then if a piece of iron were placed near to x, the iron would move toward x, for there are an indefinite number of disturbing conditions that would prevent such movement. (Suppose, for example, that object y, which is more strongly magnetic than x, is also near the piece of iron.) Tests of whether something is magnetic must be qualified by an "other things being equal" clause.

Moreover, terms like "diatomic molecule" cannot plausibly be construed in this way. It is difficult to conceive of a physical operation and its result that are coextensive with a diatomic molecule, even when an "other things being equal" clause is understood. This does not mean, of course, that one cannot assert that something is a diatomic molecule on the basis of tests. But these tests are indirect; they are based on a whole complex of assumptions and theories. To suppose that some operation and its results — other things being equal — allow one to assert that something is a diatomic molecule is much too simple.

We can conclude that Bridgman's operationist program, if interpreted in terms of extensional reportive definitions, is also impossible; it is impossible to follow MR_1 if extensive reportive definitions are intended.

A more plausible interpretation is to suppose that operational definitions are rational reconstructions of scientific terms. MR_1 then could be interpreted as an imperative directing scientists to rationally reconstruct all scientific terms in terms of physical operations and the results of these operations. The critical question is whether such a program would be beneficial for science. There is good reason to suppose that, on balance, it would not be. This is not to say that some benefits would not result, but rather that the losses and disadvantages would seem to outweigh the advantages.

Let us consider analytic rational reconstructions first. The benefits of construing (2') as the definiens of (1') should be obvious. So construed, (1') would have a clear experimental sense. Since Bridgman seems to suppose that the property specified by the definiens is logically necessary and sufficient for the application of the definiendum, there would be a clear and certain means of finding out whether something was magnetic in the sense of iron-attracting magnetic. This clarity and relative certainty would be a definite advantage.

However, the price one would have to pay is high. One would be forced into admitting that "magnetic" in (1) has a different sense than in (1'). Moreover, every different operation and its results would create a different sense of "magnetic." The same would be true for all terms in physics, for "temperature," "length," "acid," and so on. For example, "tactual length" defined by the laying off of measuring rods and "optical length" defined by optical triangulations would have to be distinguished.

Strictly followed, MR_1 would thus result in an endlessly complex scientific language, since, unless these different senses were noted in scientific language by different terms, e.g., $length_1$ or tactual length, and $length_2$ or optical length, scientific language would be hopelessly ambiguous. However, this proliferation of scientific terms would defeat one of the principle purposes of scientific inquiry: to provide a simple unifying theory. In short, following the operationist program where that is construed in terms of providing analytic rational reconstructions of all scientific terms would be tantamount to eliminating scientific theories as we usually know them.

Could operational definitions plausibly be construed as extensional rational reconstructions? Again, to do so would be to sacrifice a great deal more than most people would be willing to sacrifice. Supposing that all scientific terms could be rationally reconstructed in this way, the failure of the test specified in a definiens would mean that the definiendum did not apply. For example, if (2) were false, then (1) would be also, i.e., x would not be magnetic. But it would be unfruitful to reconstruct scientific terms in this way. In science we surely want to provide that the failure of a test for something's being magnetic is compatible with the object's being magnetic. As we have seen, tests are relevant only against a background of assumptions and theories. Thus a test may fail because of an incorrect auxiliary hypothesis, rather than because the definiendum does not apply. This sort of rational reconstruction does not preserve the systematic and provisional character of science.[17]

Despite the failure of the operationist program as interpreted above, the effect of operationism on science has not been completely negative; indeed it has had great positive value. The operationist approach has forced scientists to think of ways of clarifying and testing scientific theories. In practice, operationism has meant not so much the attempt to rationally reconstruct *each term* in science in terms of physical operations and the results of these operations, but the attempt to provide clarification of and give empirical significance to *theories*. This has been achieved by formulating and reformulating theories in such a way that they have test implications of a certain form. In other words, the operationist program has urged scientists to rationally reconstruct a theory's terms so that the definiens of each term had the form:

If operation O were performed, then result R would occur.

As we have seen, this program was not acceptable. However, in actual practice scientists reconstructed the theory — not necessarily its individual terms — in such a way that the theory plus auxiliary hypotheses had test implications of the form:

If operation O were performed, then result R would occur.

It is important to see that a theory plus auxiliary hypotheses might have test implications of this form without any of the definiens of the theory's individual terms having this form. As we saw in Chapter One, it is sometimes necessary to rationally reconstruct many of the terms in a theory before it has test implications. Indeed, we suggested that the structure of some testing situations can be understood as follows:

If the hypothesis and background assumptions A are true, given a certain reconstruction, then a certain test implication is true.
But the test implication is not true.

∴ Either the hypothesis is not true, or one or more of the background assumptions is not true, given the reconstruction.

This suggests a new way of interpreting the operationist program, one that is more in keeping with actual scientific practice and is free from the problems of MR_1:

MR_2 Construct or reconstruct theories and auxiliary hypotheses so that they have test implications in terms of physical operations and the results of these operations.

Although MR_2 is surely an improvement over MR_1, certain cautions must be noted. First, MR_2 suggests that the test implications of all theories in science should be in terms of physical operations. This in turn may be interpreted to mean that all theories must be tested experimentally, by actual physical manipulations of variables. As we saw in Chapter One, this is not necessary for theory testing; for example, theories in astronomy are not generally tested this way. Secondly, it may not always be fruitful at an early stage of theory development to try to construct a theory that is experimentally testable, even if it is fruitful at some later stage. Such an attempt might hinder the creative development and formulation of the theory. It might be better to wait and let the theory develop naturally before beginning experimental translation.

OPERATIONISM AND SCIENCE EDUCATION

Although the operationist program as specified by MR_1 is undesirable, it does not follow that this program might not be useful, with appropriate modifications, in the context of science education. As we have seen, science educators need not in all contexts be restricted to what is correct in the scientific context. So it might be very fruitful for science teachers to teach new terms operationally, even if this is not what these terms mean in science and even if a rational reconstruction of these terms along operational lines is unwise. The science teacher must adjust his pedagogy to his student's development and capacity, and not entirely to scientific correctness.

This pedagogical attitude is especially appropriate in the light of recent findings in child psychology. The work of Piaget and others suggests that, for young children, physical concepts are closely connected with actual behavior, i.e., operations, and only later become internalized. In a sense, as Mays points out, Bridgman's program reverses the natural order of psychological development:

> Though Piaget would readily admit that the concept of length, for example, started off in the child as a group of behavioral activities such as comparing, ordering, etc., he would, however, hold that in the case of adult intelligence these activities have now obtained a symbolic expression in the form of mental operations, thus obviating the need for their overt performance. Indeed, the very value of a con-

cept as a mental construct lies precisely in its economy, in its power of compressing a large amount of information in a restricted space. Bridgman's approach . . . seems to invert the natural order of mental development from external action to mental operation.[18]

It would seem, then, that for young children at least the following pedagogical rule is appropriate:

PR_1 Teach scientific terms operationally.

This could be done in at least three ways. First, stipulative definitions of scientific terms in terms of operations and the results of operations could be given. Secondly, scientific terms, e.g., length, weight, could be taught ostensively by pointing to certain physical operations and their results. Thirdly, scientific terms could be taught contextually by using these terms in sentences which describe the appropriate physical operations and results. Naturally, as the child matures, a more scientifically acceptable and correct understanding of the meaning of these terms could be approximated.

These suggestions about operational definitions in science education should be distinguished from Bruner's account of the role of operational definitions in science education. Bruner considers the possibility of isolating certain "recurrent ideas that appear in virtually all science" and teaching these ideas in a "manner that frees them from specific areas of science." One of the basic ideas Bruner mentions is "operational definitions of ideas."

Bruner's discussion of operational definition is not in accordance with what is ordinarily meant by operational definition. He says:

> With respect to the last [operational definition of ideas], for example, we do not *see* pressure or the chemical bond directly but infer it indirectly from a set of measures. So, too, body temperature. So, too, sadness in another person.[19]

It is plausible from this quotation to assume that Bruner is not talking about the operational definition of pressure or chemical bond at all, but rather about the empirical indications of pressure or chemical bond, that is, the evidence one uses to infer that there is pressure or chemical bond.

Bruner goes on to consider the feasibility of teaching the idea that we infer the existence of something from certain empirical indicators. He suggests that it be taught "with a variety of concrete illustrations" in the early grades "in order to give the child a better basis for understanding their specific representation in various special disciplines later." Bruner's pedagogical recommendation, therefore, can perhaps be better understood not in terms of PR_1, but in terms of a special pedagogical application of MR_2.

In any case, MR_2 does suggest certain approaches to science education. It may be good pedagogical practice to teach science in terms of the way scientific hypotheses are tested and, in particular, to teach it in terms of the concrete manipulations involved in testing hypotheses. Students should be encouraged to ask and answer questions like "What would I have to do to get information to test this hypothesis?" "How could this hypothesis be interpreted so that a result of my action would count for or against it?" Such emphasis would fit in well with recent emphases on the inquiry approach to science education. The present suggestion, then, could be formulated by the following pedagogical rule:

> PR_2 Teach science in terms of the concrete physical procedures used to test hypotheses.

This rule should be followed, however, with caution and with certain qualifications, lest teaching science in this way encourage science teachers to neglect other aspects and dimensions of science. PR_2 stresses the context of testing; however, as we have seen, the context of generation is also part of scientific inquiry and should not be neglected in science education. Moreover, the emphasis in PR_2 is not only on the context of testing, but on concrete physical procedures: abstract mathematical deductions and complex chains of logical argument are often needed to test a hypothesis, and these should not be neglected in science education.

Whether PR_2 is especially suited to a general science course, as Bruner suggests, is unclear. There is no *a priori* reason why this approach might not be used in introductory science courses in biology, chemistry and so on. Much would depend on the maturity and cognitive development of the students. However, it may happen that students who take general science courses have a cognitive development that is especially suited to thinking in terms of concrete physical procedures. This is something educational research must decide.

•

CHAPTER FOUR

Observation

•

We have seen how Mr. Soames, the social studies textbook writer, has become interested in participant observation, an important method of observation in the social sciences. Although participant observation is not used in the natural sciences, it is generally agreed that observation of some sort is used in all sciences. It is also generally agreed that, in the teaching of science, observation is crucial. Thus science teachers teach their students how to make and report observations, and that certain scientists have made observations that support or refute certain theories. They urge their students to observe certain things in classroom demonstrations and on field trips. What is unclear or debatable is the way in which observation is important in science and science education.

In this chapter we will consider the importance of observation in science. First we will try to clarify what observation is. After an analysis of the concept of observation, we will consider the uses of observation in the context of generation and the context of testing. Consideration of the context of testing will lead to the standard objectivist account of testing theories via observation, and to the recent subjectivist attack on this traditional account. We will evaluate this subjectivist account and show that the standard objectivist view can be defended.

The general discussion of observation in science will prepare the way for a discussion of observation in science education. We will show how the distinctions and insights derived from a consideration of observation in science provide illumination and suggestions for science textbook writers, researchers in science education, and science teachers.

After the general discussion of observation in science and science education, we will focus our attention on a particular method of observation used in the social sciences, participant observation. We will argue that, although the widespread use of participant observation may not be methodologically justified in the social sciences, its use in science education has yet to be explored or appreciated.

THE CONCEPT OF OBSERVATION

The term "observe" is used in many different senses. Indeed, sometimes when we speak of someone observing something, we are not referring to his sense perception at all. Thus we say that Jones observed his twenty-fifth wedding anniversary. What may be meant is merely that Jones celebrated his twenty-fifth wedding anniversary. However, this use of "observe" is not important for scientific or philosophical purposes and will be ignored here. We will concentrate on senses of "observe" that are relevant to sense perception.

However, the typical senses of "observe" found in scientific contexts do not refer to all sense perception. "Observe" usually refers to visual perception; it is slightly strange to speak of observing something by touching it or hearing it.[1] One can perhaps broaden the term "observe" to include other sense modalities, but such a use will not be adopted here. For convenience we will restrict the term "observe" to visual perception.

Furthermore, although observing refers to visual perception, to observe something is not the same as to see it. To speak of someone observing something suggests that the person is paying close attention to some aspect of it, that he is giving the object close visual scrutiny. However, to speak of someone seeing something does not usually suggest close scrutiny or attention.[2]

To observe something, then, in a sense of "observe" relevant to scientific contexts is to visually attend to some properties or features of it. The "something" might be an event, a process, or an entity.

However, sometimes more than this minimal sense is involved. To see this, consider the following two expressions:

(1) Jones observed John's measles symptoms.
(2) Jones observed that John had measles symptoms.

In some contexts (1) and (2) might mean the same thing, but this is not typical. Ordinarily (2) seems to imply more than (1). In (1) one is merely saying that Jones is visually attending to a property of John, namely, his measles symptoms. However, one is usually not saying anything about Jones' belief or other cognitive states. Thus Jones might not believe that John had measles symptoms and (1) might still be true. Jones might not have any belief about what he is attending to or he might mistakenly believe that the symptoms are chicken pox symptoms.

However, typically in (2) something more is at issue. Belief and other cognitive conditions of the observer seem to be relevant. Let us call the first sense of "observe" — the sense of "observe" usually manifested in expressions like (1) — the *noncognitive sense*. The sense of "observe" typically found in (2), which we may call the *cognitive sense*, is also found in expressions with a form similar to (2), namely:

Person *P* observed that *x* is *Q*.

For example, "He observed that the electron passed through the cloud chamber," and "He observed that Abuta, the medicine man, was performing the ascension ceremony." Both seem to imply more about the observer and his condition than that he is visually attending to certain properties of certain objects. We will first consider one type of cognitive observing which we will call *primary cognitive observing*.[3]

In primary cognitive observation, one is usually attending to the property or feature of the object one is said to observe, e.g., John's measles symptoms, as well as being in a certain cognitive state. One must believe that John has measles symptoms. This belief must be based on that which one is visually attending to. To the observer John looks like he has measles symptoms. (To put it in the language of phenomenological psychology, that John has measles symptoms is *phenomenologically immediate*.)

However, not all cognitive observing is primary cognitive observing. People, especially scientists, are said to observe properties of objects they are not actually visually attending to. Because of certain known connections of the property to which they are attending to other properties of the same object or different object, they are said to observe these other properties. Thus a doctor observes that John has measles, although what he is attending to is the measles symptoms of John; a physicist observes that an electron passes through a cloud chamber, although what he is attending to is a white track in the cloud chamber; a field anthropologist observes that Abuta has as one of his functions the preserving of tribal harmony, although he is attending to Abuta's performing certain ceremonies.

In all of these cases the property of the object actually attended to is associated with the property of the object said to be observed: symptoms of measles are caused by the measles virus, the track in the cloud chamber is produced by the electron, performing certain ceremonies brings about tribal harmony, and so on. Knowing that this association holds gives the observer a justification for believing that a certain property occurs on the basis of his primary cognitive observing that another property holds. Furthermore, when this knowledge is well integrated and ingrained into the observer's cognitive makeup, it appears to the observer that the object has this property; his belief is not based on any process of discursive reasoning. It is just as phenomologically immediate to a trained physicist that an electron is passing through the cloud chamber as that there are white tracks in the cloud chamber. We will call this type of cognitive observing *secondary cognitive observing.*[4]

It is important to realize that in both primary and secondary observing there is no conscious inference or interpretation. What is observed is directly and immediately given to the observer in the phenomenological sense of immediacy, i.e., if someone observes that *x* is *Q*, then it visually appears to him that *x* is *Q*. The reader should not be led to suppose that secondary observing is based explicitly on discursive reasoning, while primary observing is not. In neither case is such reasoning involved.

THE INFLUENCE OF THEORY ON WHAT IS OBSERVED

The theoretical background and training of the observer greatly affects what he observes and can observe. It is clear that in primary cognitive observing much of what can be observed is a function of the background information and training of the observer. A person without appropriate information and training could not observe that John had measles symptoms. This dependence of what can be observed on background information and training is common in science. A field anthropologist can observe that Abuta is performing the ascension ceremony; a layman attending to the same properties of Abuta, because of his lack of information and training, cannot observe this. A chemist can observe that a certain liquid is mercury; a layman attending to the same properties of the liquid cannot.

The differences between what a scientist and a layman can observe are even more striking in secondary cognitive observing. The scientist's knowledge of certain associations between different properties expands enormously what he can observe, relative to the layman. For example, the physicist, because of his knowledge and training, while attending to tracks in a cloud chamber, can observe that an electron is passing through the cloud chamber, although he is not attending to the electron; a layman, because of his lack of knowledge and training, while attending to tracks, does not observe that the electron is passing through the cloud chamber. A physicist attending to the position of weights on a balance scale can observe that one weight has more mass than the other; the layman attending to the same thing cannot.

Furthermore, as more of these associations — embodied in scientific laws, theories, principles, and the like — are learned, the range of what can be observed increases. Indeed, in the secondary cognitive sense of observe there is no *a priori* limit to what can be observed. The only limit is the lack of knowledge of certain connections and associations between properties of objects and the assimilation of this knowledge by the observer.

We have argued so far that a trained observer with certain knowledge and training can observe things that a person without this knowledge and training cannot observe. In particular, knowledge and training affect what one can observe in the primary and secondary cognitive types of observing found in science.

However, the background and training of the scientist also influence what he noncognitively observes. A person's background will influence what properties he visually attends to in a particular object, or indeed whether he attends to any properties of the object at all. A doctor, because of his knowledge, may visually attend to the white spots on the inside of his patient's mouth as well as to the red spots on his face and chest, while the members of the patient's family, because of their lack of knowledge, may attend only to the red spots on the patient's face and chest. In this case the background of the scientist influences him to attend to different properties of the object from those the layman attends to.

A layman walking through familiar country may fail to notice, let alone attend to, certain features of a rock formation, glacial deposits and the like, that a geologist attends to closely. In this case, the theoretical background of a scientist leads him to observe noncognitively objects which the layman, because of his lack of theoretical background, does not observe at all.

There is a difference in the influence of a person's theoretical background and training on cognitive observing and on noncognitive observing. In cognitive observing the scientist can observe what the layman cannot; in noncognitive observing the scientist tends to observe what the layman, although he could observe it, does not observe. For example, members of the patient's family could attend to the white spots in the patient's mouth, but because of their lack of training do not think to do so; however, members of the patient's family could not observe that he had measles, given their lack of training.

The above examples of the influence of theoretical background on noncognitive observing, as well as our discussion of the effect of theoretical background on cognitive observing, may have given the impression that the background and training of a scientist are always an advantage in observing. However, the theoretical background and training of a scientist may influence him to fail to noncognitively observe what a person without this theoretical background and training might noncognitively observe.

Thus a psychoanalyst, because of his theoretical orientation, may attend to certain behavior of his patient which is not attended to by nonpsychoanalysts. But this same orientation may lead him to overlook behavior which is attended to closely by psychologists with other theoretical orientations, and which might be just as important in understanding the patient as the former type of behavior.

Not only can a scientist's theoretical orientation influence him to neglect certain properties which it might be important for him to attend to, but it can also cause him to make certain errors in judgment about what he is attending to. His theoretical orientation may influence him to believe things that are not true. Indeed, it is not unknown for a particular theoretical orientation of a scientist to blind him to evidence which is negative relative to that theoretical orientation.

For example, suppose that according to a certain theory strongly held by person P all A's are B. Suppose that P attends to object x that is A and not B. The existence of such an object should provide a counter-example to this theory. However, P's strong commitment to this theory might influence him to believe that x is A and C. But if x is A and C, this would not count against his strongly held belief. Thus his theoretical orientation influences him to make an observational judgment that is not incompatible with his theoretical orientation.

An experimental analogue of this phenomenon has been produced in the psychological laboratory. Bruner and Postman have shown how subjects' perceptions of an incongruity emerge with great difficulty because of prior expectations.[5] Subjects in the experiment were asked to identify in a short exposure a series of playing cards. Some of the cards were abnormal, e.g., black four of hearts. All subjects found it easy to identify the normal cards, but difficult to identify the abnormal cards as abnormal. The tendency was to identify the abnormal cards as normal, e.g., the black four of hearts as a black four of spades. Only after long exposure were the abnormal cards correctly identified by the majority of the subjects, and some of the subjects were unable to make correct identification even after repeated exposure.

This experiment illustrates well the point that one's expectations may influence one to overlook exceptions to them. The same sort of phenomenon happens in science,[6] and we will consider the methodological problems connected with this phenomenon in a moment.

THE USES OF OBSERVATION IN SCIENCE

OBSERVATION IN THE CONTEXT OF GENERATION

It has often been said that modern science is based on observation. But in what sense is this true?

It may be supposed that all scientific theories and hypotheses arise from observation and generalization from this observation. But this view is simply not in accordance with the historical facts: some scientific theories are generated in other ways, e.g., by dreams, inspiration, and so on.

It may be supposed that, although scientific theories are not always generated from observations and generalization from this observation, they *should* be so generated. But, as we have already seen, how a theory is generated is irrelevant to its truth; the crucial question is how the theory stands up under test. To be sure, some processes of generating scientific theories may be preferable to others because they lead to more fruitful, testable, or true hypotheses. But there is no *a priori* reason for thinking that observation and generalization from this observation is to be preferred on these grounds, and there is little evidence to show that they lead in all contexts to hypotheses that are more fruitful, testable, or true.

Indeed, it can be argued that observation and generalization from this observation cannot generate particular kinds of theories. Generalization from what is observed by a person can only yield theories about what is *observable* to that person. Thus, if a theory is about unobservable entities, it cannot be generated by generalization from what is observed.[7]

There is no reason to prefer observation and generalization from observation as a method of generating hypotheses in all contexts, and indeed there is some reason to suppose that in some contexts this would not be a fruitful method. However, some methodologists have gone too far. They have argued that this method cannot possibly generate new hypotheses that are fruitful, testable, true, and so on.[8] But surely this is mistaken. Whether observation and generalization from observation is a good way of generating new hypotheses is not something that can be decided by *a priori* reasoning; the question must be decided empirically.

It is instructive, however, to consider a possible reason for this extreme view. One might argue as follows: Observation is not the pure and simple affair that most people have thought. Indeed, observation is largely influenced by one's perspective, theoretical commitments, expectations, and so on. It is selective: what one notices or focuses on in perception is influenced by certain ideas and assumptions. Now consider a person who observes that in all the cases he has investigated sodium salt burns yellow, and who seems to be generalizing from his observations that all sodium salt burns yellow. The thesis that he has generated a new hypothesis is, however, an illusion. His observations are necessarily influenced by some ideas or presumptions. In his case, this could only be the idea that all sodium salt burns yellow. This assumption has guided his investigation. Thus no new hypothesis is generated by observation and generalization from this observation, since the hypothesis allegedly generated was implicit all along in the selective process of observation.

However, simply because one's observations of sodium salt burning yellow have been guided by *some* presumptions, it does not follow that they are guided by the conclusion that all sodium salt burns yellow, the very conclusion of the generalization. For example, a person doing his investigation might be guided by the assumption that all sodium salt, because of chemical makeup M, burns either yellow or orange or red, but not blue or green, yet this assumption is surely different from the generalization he makes on the basis of observation guided by it. Moreover, even if the person were guided in his observation by the tacit assumption that all sodium salt burns yellow, the generalization to all sodium salt burns yellow would make this presumption explicit; it would bring this assumption to light. Thus, although no new hypothesis would have been generated in one sense, in another sense a new hypothesis would have been generated, would have been made explicit.

Although it is possible to generate hypotheses by observation and generalization from this observation in certain contexts, this is not the only, perhaps not even the most common, role that observation plays in the generation of hypotheses. Several other roles should be mentioned.

Hypothesis generation typically begins when a problem is observed; the observation of the problem starts the creative wheels turning. However, as we have seen, observation is often hindered by theoretical bias. A scientific investigator who makes a particular set of assumptions often finds it difficult to see problems in his system; his theoretical biases blind him to his theory's shortcomings. On the other hand, one's theoretical commitments may make one sensitive to certain problems; problems that are easily spotted by one who follows a particular theory may be far from obvious to one who does not. Nevertheless, observation of a problem, whether helped or hindered by theoretical commitment, is an important first step in theory generation.

Sometimes even after one observes some problem the problem resists solution until the situation in which it is embedded is observed in a new way. This refocusing may happen in a twinkling of an eye.[9] In any case, the changed vision enables one to conceptualize the problem differently and develop new hypotheses in answering the problem. This phenomenon of refocusing one's visual field, studied by Gestalt psychologists, is still not well understood.[10] But this at least seems clear: among the factors furthering or hindering the refocusing are one's theoretical commitments.

Recognizing analogies or observing important similarities between classes of phenomena is often a crucial step in the generation of hypotheses. Observing the analogy between electricity and water may enable an investigator to formulate a hypothesis about electrical flow that is in certain important respects similar to well-established hypotheses about the flow of water.[11] This ability to see analogies may be furthered or hindered by one's theoretical orientation.

However, the well-established fact of the influence of theory on observation has led some methodologists to make some misleading claims. Charles Darwin, for example, argued that every useful observation is made for or against some hypothesis.[12] Now if Darwin's point were just to emphasize the influence of theory in observation, it could be accepted. However, his statement may wrongly suggest that the motivation for observation in the context of generation is always the same, to refute or establish some hypothesis. But one may observe something for any number of reasons; motivation for observation differs from context to context and person to person. A scientist may attend to something that is fascinating and interesting to him; as a result he may make an important discovery. As we have seen, what the scientist takes as fascinating and interesting, what strikes him as curious, is influenced at least in part by his theoretical orientation. But it is one thing to admit this and another thing to say that every useful observation is made for or against some hypothesis.

It may be argued that, in the context of testing, every useful observation *is* made for or against some hypothesis. Indeed, it may be said that this is the barest tautology; in testing one has a hypothesis and one is making observations that one believes either refute or support the hypothesis. The correct motivation for observation in the context of testing is the same. However, it would be a misunderstanding of the context of testing—no less than of the context of generation—to suppose that some particular psychological motivation is a necessary condition for useful observation. An observation made for any reason may still be considered an important observation in the context of testing so long as it is relevant to the testing of a given hypothesis. In the same way, an observation made for any reason may be considered an observation in the context of generation if it leads to the generation of a hypothesis; the observation in order to be useful need not be motivated in any particular way.

Of course, one might adopt a policy that in order to generate a hypothesis all observations must be made for or against some hypothesis. In short, one might advocate that scientists always be motivated in certain ways in their observations. But there is no reason to suppose that such a policy would be feasible and there are psychological and logical reasons to doubt that it would be. First, such a policy would be extremely difficult to follow and might well hinder the creative process, which in scientific inquiry often works in subtle and subliminal ways that are not allowed for in the present recommendation. Secondly, if every observation used in generating a hypothesis must be made for or against some particular hypothesis, one might well ask how that particular hypothesis was generated. Must that hypothesis also have been generated by some previous observation for or against some other particular hypothesis? If so, we seem to have an infinite regress; if not, then it is difficult to say why one should make observations for or against some particular hypothesis in one case and not in the other.

Although the policy that observations in the context of generation should be motivated in one particular way is inadvisable, this is not to suggest that observation in the context of generation should be unmotivated; moreover, it is surely not to suggest that observation should be naive. Observation, as we have construed it, is an activity that can be done well or badly, within the context of well-established theory or within the context of ill-informed ideas and prejudices. Moreover, it is an activity that is learned and can be improved upon. The idea that the best scientific observation is naive like a child's, and that to become a good scientific observer one must unlearn all the ideas one has previously learned and return to the pristine purity of infancy, has little merit. Becoming a good scientific observer involves great sophistication and skill. The cure for bad scientific observation is not naive observation, but more training and sophistication. How this sophistication may best be accomplished we will consider later.

OBSERVATION IN THE CONTEXT OF TESTING

The role of observation in the testing of scientific theory is fundamental. If there is a clear sense in which science is based upon observation it is here. Scientific theories are primarily tested against observation and accepted, rejected, or modified mainly because of the observational data. Observation is thus generally considered to be the touchstone of objectivity in science; it seems to be primarily observation that provides an independent standard for the evaluation of theories and hypotheses. If it were not for observation, there would be little reason for choosing between scientific theories and fictional accounts, between science and pseudoscience, between warranted assertions and fanciful hopes.

Observation and Certainty. However, to say that observation is fundamental to the testing of scientific theories is not to say that observation is infallible, or that an observational report cannot be changed or rejected in the light of other observations or theoretical pressures. A necessary condition for utilizing observation as an independent standard for evaluation is not that observation is certain. We shall consider in a moment what observation must be for it to serve as an independent standard for evaluation.

In any case, observation clearly cannot be maintained as infallible or certain. The existence of perceptual illusion, hallucinations, and other less dramatic perceptual errors proves that people can be deceived by their senses. Those who wish to maintain the certainty of observation are forced back to a new level of observational reports. This is accomplished through the argument that, although observational reports about physical facts can be mistaken, observational reports about one's own sense impressions cannot be. Consider the following observational report of Jones:

(1) The liquid in the test tube is light red.

This observational report could be mistaken. For example, the test tube may be painted light red three quarters of the way up so that under certain conditions the illusion that it is filled with light-red liquid will be given. However, suppose the observation report is not about a physical object but about one's sense impressions, that is, about what appears to the observer. For example, Jones might report:

(2) It seems to me that this test tube is filled with light-red liquid.

In the case of (2) there is no need to worry about illusion, for the report is not about a physical fact but about what is appearing to the observer. Whether what is appearing has any relation to physical facts is not in question.

Advocates of certainty in scientific observation have often admitted that observational reports of physical facts are fallible. They have insisted, however, that observational reports of sense impressions are infallible. Furthermore, they have maintained that, because of this certainty, the observational basis of science should be observation of sense impressions.

We may well ask whether observational reports of sense impressions are infallible, and even if they are, whether such reports should be used as an observational basis of science. The answer to the first question is no. Observational reports of sense impressions can be in error. Such mistakes can occur in several ways.[13] First, linguistic errors of various sorts are possible. A person can use a wrong word in his observational report. He may have the sense impression of a test tube filled with light-yellow liquid and may mistakenly use the word "light-red," either because of a slip of the tongue or because of ignorance of the language. Secondly, non-linguistic errors are possible. A person may simply misdescribe his sense impressions, not out of linguistic confusion or error, but because of inattention, lack of care in describing what is appearing to him, or the complexity of his sense impressions. The answer to the second question is also no. Sense impressions—whether the reports about them are fallible or infallible—vary from person to person; they are about private phenomena. But in order to maintain the objectivity of science, the language of science—whatever else it must be—must be public.

We may conclude that the move to the level of sense impressions in order to provide a bedrock for science is ill conceived. Observational reports of sense impressions are not certain, and even if they were, there are good methodological reasons for not using them. The public, although less than certain, nature of observational reports of physical facts must provide the basis for science.

The Acceptance of Observational Reports. As we have seen, observation is not certain. But, it may be asked, if observation is not certain, how can it have any relevance to theory testing and evaluation, how can it be an independent standard of evaluation? The answer is this: Observation made under certain appropriate circumstances does have a *prima facie* claim to be regarded as correct. The burden of proof, under these circumstances, is on the person who rejects or modifies an observational report.

The circumstances under which observational reports should be regarded as *prima facie* correct depend on the situation and the type of observation at issue.[14] The state of mind of the observer while he was making the observation, e.g., whether he was under the influence of drugs, would clearly be crucial in most—if not all—circumstances. The past history of the observer should also be taken into account in most circumstances. Thus, for example, an observation made by a person who suffered from hallucinations surely would have no *prima facie* claim to acceptance. Moreover, the physical circumstances under which the observation is made are crucial. An observational report of what is seen under poor lighting conditions would usually have no *prima facie* claim to acceptance. In situations in which certain tools or instruments are used, e.g., telescope, eyeglasses, or X rays, the acceptability is a function of the reliability of the tools used. This reliability in turn depends on the theory in accordance with which the tools were constructed and conceived.

This last point suggests that the circumstances under which observations should be taken as *prima facie* acceptable are subject to change. This change will sometimes depend on changes in evidence and theory. For example, on the basis of our present evidence and theory, an observation report made through a telescope under normal conditions of the positions of a star would be *prima facie* acceptable. However, the telescope is based on certain principles of optics which are subject to change. If a principle were rejected, this might in turn affect the acceptability of the observational report made by use of the telescope.

In any given context, an observational report which would ordinarily be considered *prima facie* correct may be rejected if there are good reasons for doing so. One good reason might be the isolated nature of the report. For example, suppose an observational report were received that a man lived without food or water for six months. The observation was made, as far as one could tell, under standard conditions; the observer was not drugged, he was reliable, and so on. Nevertheless, one might well reject such a report despite its *prima facie* correctness if this were the only report received, if it were not corroborated by independent observers.

Two kinds of rejection of a *prima facie* correct observation report are possible: the report may be considered false, or one might suspend judgment on the report.

Rejection of the report may be combined with its replacement by a new observational report couched in different language. This new report can then be rejected in turn in either of the above senses or accepted. If it is rejected, it may be replaced by yet another observational report couched in still different language, and so on.

For example, consider an observational report of a piece of copper that did not conduct electricity. This report was made under standard conditions by a competent observer in a normal state of mind. Thus the report has a *prima facie* claim to acceptance. But suppose that there is good reason to reject the report. One might say that the report is false in that the person did not observe what he claimed to observe, or one might say that at the present time we can make no judgment about the truth or falsity of the report. Either of these moves could be combined with a new description of what the person claimed to observe. The person might be willing to describe what he observed as a reddish metal with atomic weight 29, rather than as copper. The new observational report that this piece of reddish metal, etc., did not conduct electricity would be considered in the same way: it might be rejected for good reasons in either sense of "rejection," or it might be stated by the observer in still different language and reevaluated.

OBSERVATION AND SCIENTIFIC OBJECTIVITY

The approach suggested above, namely, that observation made under certain conditions should be considered as *prima facie* acceptable, is compatible with the objectivity of scientific theory testing via observation. For observation so considered can be interpreted, as we have interpreted it, as providing an independent standard for evaluation of scientific theories. A *prima facie* acceptable observational report, although it can be rejected for good reasons, need not be so rejected. If there is no good reason for rejecting it, such a report might well conflict with some accepted theory. The theory in this case would have to give way; it would have to be modified or rejected. Thus observation can conflict with a theory and provide an independent standard for evaluating the theory. On the other hand, if an observational report is deducible from the theory, the report would provide some independent support for the theory (at least according to the confirmation approach).

In recent years the objectivist view of theory testing via observation has come under serious attack. The attack centers on the influence of theory on observation.[15] Earlier in this chapter we saw how theory can influence observation. Critics of the objectivist approach—methodologists we shall call *subjectivists*—have used the facts of the influence of theory on observation to undermine the objectivist position. The attack has centered on two aspects of the objectivist view: the independent standard of observation for evaluating theories and the possibility of empirical comparability of rival theories in a domain. The subjectivists have argued that observation is not really an independent standard of theory evaluation because of the influence of theory on observation. They have argued also that, because of different observational languages relative to different theories, it is impossible to compare empirically rival theories in a domain. Let us consider these two arguments in more detail.

It has been argued by philosophers like T. S. Kuhn that if a theory influences observation—as psychology and the history of science show that it does—the support provided a theory by observation is specious.[16] The independent standard of observation, so it is argued, is thus an illusion. For example, an observer who is a Newtonian—and thus committed to Newton's theory—is influenced by his commitments in making observations in testing Newton's theory. His observations will naturally conform to the theory and thus provide specious support for the theory. His evidential support would be circular. Thus the observations of a Newtonian could not provide any independent standard of evaluation for Newton's theory.

It has been argued also that rival theories in a given domain cannot conflict empirically. The observational language of advocates of one theory will be different from the observational languages of advocates of the other theory. There is, it is argued, no neutral observation to test the theories against. All observational languages are "theory laden."[17] The observational language used to test each theory is relative to the theory tested. For example, suppose that behavior theory and psychoanalysis were rivals in a given domain. Behavior theory might predict that a person under a certain condition C would exhibit avoidance behavior. Let us call this prediction Ob. Psychoanalytic theory might predict that the same person under the same condition will regress to an earlier level of psychosexual development. Let us call this prediction Op. These theories, according to the subjectivist, because of their different observational languages are not comparable since their predictions are neither the same nor in conflict.

Let us consider these two arguments in turn. We will call the first argument the *argument from circularity* and the second argument the *argument from incomparability*.

The first thing to notice about the argument from circularity is an internal problem. In order to support his case, the subjectivist appeals to evidence from the history of science and psychology. He seems to argue that his evidence shows that the confirmation of a theory is specious and circular. But such an argument involves the following problem: There does not seem to be any reason to suppose that the subjectivist position itself is immune to specious and circular support. Thus it could be argued that the evidence cited from the history of science and psychology only supports the subjectivist position because it is construed in terms of a certain theory, so that the support of the subjectivist position is itself circular and specious. Put somewhat paradoxically: If the subjectivist position is true, we have no good reason to believe it is. Of course, the subjectivist may wish to argue that his position is exempt from circular support, but it is difficult to see how such an exemption could be justified.

This internal problem suggests that there is something very wrong with the subjectivist position. Nevertheless, the initial plausibility of the position seems to remain. Indeed, its plausibility seems to rest on facts of language, psychology, and history which a person who advocates the objectivist view of science surely must admit. Unless the advocate of the standard view can find a flaw in the subjectivist argument, he will be forced to accept it despite this internal problem. Pointing out this internal problem, then, must be supplemented with further analysis and argument to refute the attack on the objectivity of science.

This further analysis and argument is provided by Israel Scheffler in his book *Science and Subjectivity*.[18] Scheffler makes a distinction between (1) the categories in which observational reports are made and (2) the observational reports themselves.

(1) It is true that the categories used in observational reports are relative to certain theories. Observational reports in this sense are theory dependent and necessarily so. Let us call this influence of theory on observation *category influence*. Consider the following observational reports:

    (a) An electron passed through the cloud chamber.
    (b) Jones has the measles.
    (c) Abuta, the medicine man, is performing the ascension ceremony.

The categories reflected in the language of these reports — electron, cloud chamber, medicine man, ascension ceremony — are all relative to certain theoretical frameworks and the orientations of the observers. Let us grant that the theoretical orientations of the observers predetermine how any observation will be categorized in the observational report used to test these theories. However, these theories do not predetermine the *particular* observational report that is made. Thus the theoretical categories of doctors may predetermine that any observation by these doctors used to test the theories of medical science be stated in certain medical categories. However, the theoretical categories of medical science do not predetermine that a doctor will observe that Jones has the measles.

To be sure, medical science may predict that Jones will have the measles. But the truth of this prediction is in no way guaranteed by the classification scheme found in medicine. What may be guaranteed is that Jones will be described in medical terms. It is well to remember that theories of science are not just classification schemes: theories make definite statements about what will be observed. Observation reports used to test the theory may have to be formulated in terms of the theory, but they do *not* have to be in accord with the premises.

Thus observational reports, although stated in terms of the categories of the theory, may well conflict with the premises of the theory. There is nothing in the theory dependency of observational categories to prevent this conflict. If this is granted, then there is nothing in category influence to undermine scientific objectivity. The support for a theory provided by an observational report is not shown to be circular by category influence, since observational reports couched in terms of the language of the theory can conflict with the premises of the theory.

So far we have assumed that the categories of a theory predetermine how the observational reports used to test the theory will be conceptualized. But this is by no means obvious. Just because observational reports used in testing a theory must be categorized in terms of *some* theory, it does not follow that these reports must be categorized in terms of the theory under test. It might be possible to make observational reports in some relatively neutral observational language, i.e., a language that is not free of all theoretical categorization, but that is not in terms of the categories of the theory under investigation. Of course, such observational reports might have to be translated into the language of the theory under investigation in order to be relevant to testing it. We will consider this possibility in a moment.

(2) The theory dependency of observational reports would seem to be a more serious matter for scientific objectivity. For if what is accepted as a true observational report is dependent on the observer's theoretical commitment to the premises of the theory, how can there be an independent standard? Theoretical commitment would in this case determine not only the categorization of an observation report, but the observational report itself. Let us call this influence of theory on observation *premise influence*.

It is necessary to distinguish several different theses about premise influence:

(1) Premise influence is strong in some cases of observation.
(2) Premise influence is strong in all cases of observation.
(3) Strong premise influence, when it is present, can be detected and overcome.
(4) Strong premise influence, when it is present, cannot be detected and overcome.

It seems clear in the light of the evidence that thesis (1) is true. The history of science and experiments like those of Bruner and Postman show this. There is no reason at all, however, to suppose that (2) is true, for if it were, advocates of a theory would never make observational reports that were in conflict with their theory; a scientist's observation would never be a shock to his expectations. But this is absurd.

Furthermore, even if (2) were true, this by itself would not undermine the objectivity of scientific theory testing via observation. For even if all observational reports were strongly influenced by the theoretical commitment of the observer, such theoretical influence might be detected and overcome. There is good reason to suppose that (4) is false, hence that (3) is true. There do seem to be means of detecting and correcting premise influence. For example, for most subjects in the Bruner and Postman experiments, repeated exposure to an incongruity was enough to bring the incongruity to light.

The possibility of detecting theoretical influence fits in well with our previous discussion of the acceptance and rejection of observational reports. As we have seen, observational reports made under standard conditions have a *prima facie* claim to acceptance. However, this claim can be defeated if some good reason is available. One good reason for rejecting a *prima facie* acceptable observational report is premise influence. That is to say, one might have good reason to suppose that an observer who makes an observational report is so strongly influenced by his commitment to certain premises of a theory that his report should be disregarded.

Despite the possibility of detecting premise influence in observation, this influence poses important practical questions: What is the best way to detect premise influence on observation? What is the best way to prevent premise influence? These problems, in turn, as we shall see, generate important questions in science education.

The argument from incomparability can be met in at least two ways. First, advocates of the two theories, e.g., a behavior theorist and a psychoanalyst, might agree on some neutral description of the predicted behavior. To be sure, such a description would not be neutral in any absolute sense; it would not be neutral with respect to all theoretical categorization, for no observational report can be. Yet this observational language could be neutral with respect to the observational categories of the rival theories at issue; it could be theoretically neutral in a relative sense.

For example, consider some relatively neutral language N, i.e., neutral with respect to the categories of psychoanalysis and behavior theory. Such a language might be ordinary nontechnical English, or perhaps the language of Lewin's field theory. The prediction Ob from behavior theory and the prediction Op from psychoanalytic theory might be translated into N. Suppose that N was the language of Lewin's field theory. Then Ob might become "the person will move away from some object of negative valence in his life space." Let us call this translated prediction On. However, Op translated into N might become "the person will move toward this object of positive valence in his life space." This prediction entails $-On$. The psychoanalyst and the behavior theorist would in such a case be making conflicting predictions, despite the differences in their observational languages.

Secondly, a relatively neutral observational language might not even be necessary. It might be possible to translate one theory into the other, e.g., behavior theory directly into psychoanalytic language or vice versa. Suppose that Op, when it is translated into behavior theory language, becomes $-Ob$. In this case, too, psychoanalysis and behavior theory would contradict one another.

The problems involved in effecting such a translation — in particular the problem of how much of the original meaning must be maintained in the translation — have been touched on in Chapter Three. But in any case, it is clear that in principle there is nothing in the fact that rival theories use different observational languages to prevent the theories from being in agreement or disagreement.

THE CHOICE OF AN OBSERVATIONAL LANGUAGE

We have already seen that there are alternative descriptions of empirical phenomena, and thus alternative observational languages. We can make observational reports of sense impressions or physical facts; moreover, within the language of physical facts many alternative observational languages are open to us. For example, human behavior can be described in psychoanalytic terms, in behavior theory terms, or in terms that are neutral to both theories. Although, as we have seen, a physicist would normally describe what he observed as an electron passing through a cloud chamber, he could describe it as a track produced by an electron, as strings of water droplets that have been condensed on gas ions or as just a long thin line.[19]

The question is which observational language should be used? Much will depend on the purposes and context of inquiry. The language of sense impressions is not suitable for science since a public language is needed and the language of sense impressions is not public. Nevertheless, although such an observational language should not be the basic language of scientific observation, it may still have a place within the framework of the language of physical facts. For example, subject S in a psychological experiment may report what he observes in sense impression language:

(1) It seems to me I see the dot of light move.

However, the psychologist recording this experiment would himself use not the language of sense impressions, but the language of physical facts:

(2) S reported, "It seems to me I see the dot of light move."

The pyschologist would use the public language of physical facts to talk *about* the subject's use of the private language of sense impressions. In this way the language of sense impressions has a use in psychology.

The choice between the observational languages of psychoanalysis and behavior theory and some other observation language would again depend on several factors. If one theory proved much better in all contexts, the observational language of the better theory should be preferred; indeed, in this case the alternative theory and observational language would and should be given up completely. On the other hand, each theory might have advantages in certain domains. In this case, the choice of language should be relative to the domain at issue. However, if we are interested in testing one theory *against* the other, then a translation into some neutral observation language may be the best way to handle the problem.

Which observational language a physicist uses to describe an experiment will depend on many factors. A beginning physicist, because of his lack of training, would not speak of observing that an electron passed through the cloud chamber. Such language would also have been out of place in the early stages of subatomic physics, when the theoretical import of such experiments were not well understood even by advanced physicists. Moreover, even a knowledgeable contemporary physicist might speak differently to different audiences, not because of *his* lack of knowledge of the state of contemporary physics, but because of the knowledge of his audience. He might describe what he observes to his beginning research assistant as tracks produced by electrons and to his beginning freshman physics class as white lines.

In sum, the choice of observational language must depend on several factors: the knowledge of the user of the language and his audience, the state of science at the time, and the purposes of investigation. From this, it does not follow that the choice of an observational language is arbitrary. On the contrary, it is a matter of judgment, and one's judgment can be unwise and even mistaken.

OBSERVATION AND SCIENCE EDUCATION

So far we have considered what observation is, how theory affects observations, the uses of observation in scientific inquiry, the objectivity of theory testing by observation, and the choice of observational languages. What relevance has this discussion for science education?

SCIENCE TEXTBOOK WRITERS

We have seen that observation and theory are closely connected. In particular, the theoretical background and training of an observer affects what he can observe. What is an interpretation for one observer, given his background, will be an observation for another observer, given his background. Unfortunately, science textbook writers seem to have too simple a view of observation. Consider, for example, the view implicit in the well-known and widely used E.S.C.P. textbook *Investigating the Earth*. A sharp distinction is made in this book between observation and interpretation. We are told that we observe that the moon moves across the sky, that we interpret it as moving around the earth and do not observe that it is doing so.[20]

From our previous discussion, however, it is clear that this view will not do. A naive observer might not observe that the moon moves across the sky at all. He might only observe that a white object moves across the blackness overhead; on the other hand, a sophisticated observer—one with proper training and theoretical background—might well observe that the moon moves around the earth.

A similar point is brought out in another example in the book. We are told that we can determine our latitude on the earth's surface by celestial observation, e.g., in the northern hemisphere by computing the angle of the North Star relative to the observer's position on the surface of the earth.[21] However, it is unclear why this procedure is called "celestial observation" and not "celestial interpretation," given the author's earlier discussion of observation and interpretation. From our earlier discussion, however, we know that observation in this case also will depend on the training and background of the observer. To the experienced navigator it may be obvious at a glance that his ship is at the equator; he will observe that the ship is at the equator. To the beginning navigator such a judgment will be based on a long and involved inference based on his observation of the position of the North Star, the angle shown in his quadrant, and the information in his navigation manual. He will not observe that the ship is at the equator; he will interpret that it is on the basis of what he observes.

Interpretation and observation are much more context bound than the writers of *Investigating the Earth* would have us believe. It is not surprising if science students reading such an account of observation and interpretation get a wrong picture of the role of observation in science. In any case, a knowledge by textbook writers of the philosophical issues connected with observation and theory would certainly improve the sophistication and clarity of their work.

Exercises might be designed to bring out the close relation between observation and knowledge and background. An exercise in observation and interpretation at the beginning of the book might be suggested again toward the end, after students have become more sophisticated observers. Students might be made aware of their changing powers of observation and the shifting line between observation and interpretation by such questions as:

> How do your observations and interpretations made in this exercise compare with your observations and interpretations made at the beginning of the first chapter?
> What can you observe now that you could not observe at the beginning?
> How can you account for this change?
> What might a trained scientist observe if he performed the same exercise that you just performed?

RESEARCHERS IN SCIENCE EDUCATION

In our discussion of objectivity and the influence of theory on observation, we distinguished between category influence and premise influence. We saw that, although category influence was pervasive and unalterable, it was innocuous, while premise influence, although a serious methodological problem, was not pervasive and not unalterable. Indeed, there is no *a priori* reason for premise influence to occur at all. Nevertheless, it does sometimes occur. A practical question is how to prevent it from occurring. This question in turn generates an important problem for researchers in science teaching. What is the best way to teach students to observe so that premise influence does not occur? Let us critically consider some possible answers.[22]

One too easy answer would be to teach students to observe without any theoretical commitment, to be completely theory free. (By "theory free" I do not mean observing without the use of categories. I simply mean observing without being committed to any premise of a theory.) This answer will not do, for the simple reason that in general it is methodologically unfruitful not to have some theoretical commitment in observation. Empirical research without theory would be blind and chaotic; theory gives direction and purpose to research and throwing out theory in order to avoid theoretical influence in observation would be like throwing out the baby with the bath water.

Nevertheless, this answer does suggest a certain ambiguity in the notion of theoretical commitment. In one sense of commitment, one is committed to a theory if one strongly believes that the premises of the theory are true, or at least approximately true. In a weaker sense of commitment, one is committed to a theory if one strongly believes that the premises of the theory are worth investigating, that they are useful guiding principles in research, that they are fruitful as "working hypotheses." In this weaker sense, no belief in the truth or approximate truth of the premises of a theory is presupposed.[23]

It might be said that science educators should teach students to become committed to theories in the weaker sense of commitment, and not in the stronger sense. However, this interesting suggestion seems too restrictive. Surely scientists and students have, in the light of powerful evidence, the right to believe to a high degree that a theory is at least approximately true. It seems strange to say that a biology student should not have a high degree of belief that the theory of evolution is at least approximately true. In the light of contemporary evidence, such a theory is surely much more than a useful guide to biological research and the biology student has every right to so regard it.

Nevertheless, this suggestion does call attention to the fact that one should not be committed to *all* theories in the strong sense of commitment, that many theories ought to be regarded by students of science only as useful research guides, as working hypotheses. Thus the science educator should see to it that his students regard theories as more than working hypotheses when they are justified in doing so, and not otherwise.

However, although the distinction between the strong and weak sense of commitment is important, it does not overcome the problem of the influence of theoretical commitment on observation. First, scientists and students of science still have the right, as we have seen, to be committed to some theories in the strong sense. This commitment may insidiously affect their observational judgment. Secondly, even when scientists are committed to a theory in the weaker sense, it is unclear how this commitment affects what they observe. One important research problem is to determine the differences between these two different types of theoretical commitment on observational judgment. One suspects that this influence is weaker with cases of commitment in the weaker sense, but this is not obvious. It needs empirical observation.

Another suggestion that seems to have some initial plausibility is that students of science should be taught a number of different theoretical approaches in a domain of research.[24] If necessary, discarded theories from the history of science should be resurrected and reexamined. Students should not only be exposed to different theoretical approaches, but should also learn to work easily with different theories, now seeing the domain from the point of view of one theory, now seeing it from the point of view of another, switching back and forth to get various theoretical perspectives and insights.

This idea could be combined with the idea of working hypotheses suggested above. Students could be exposed to varying and conflicting working hypotheses within a given domain. Indeed, students might even learn to be committed to many theories as working hypotheses. There is nothing irrational in this. Unlike commitment in the strong sense, a person is not irrational if he is committed to incompatible theories T_1, T_2, . . . T_n in a domain simply as useful research guides, as working hypotheses. Indeed, there is nothing incompatible in a student being committed to theory T_1 in the strong sense of commitment in a domain and at the same time being committed to several incompatible theories in the weak sense. Thus a student can strongly believe that the theory of evolution is true or at least approximately true, and yet with perfect consistency strongly believe that certain theories that are incompatible with the theory of evolution are useful guides to research.

The pedagogical principle of the proliferation of theories could be stated in this way: Students should learn to work with many theories as working hypotheses in a given domain of inquiry, even if commitment in the strong sense to an incompatible theory is justified in the domain.

The rationale for the proliferation of theories approach is this: The more theories one is used to working with in a given domain, the less likely it is that one will be blinded by one's commitments— in either the strong or the weak sense—to any one of them; the biases implicit in viewing the world through various theories will tend to cancel each other out. Evidence that is negative relative to theory T_1 that one would be insensitive to while using T_1 might well be recognized as such if one used T_2. The reverse might also hold. But since one would be working with both T_1 and T_2 in this approach, possible negative evidence relative to either T_1 or T_2 would not be overlooked or misjudged.

The rationale for the proliferation of theories should, however, be considered by researchers in science education as a working hypothesis, a hypothesis that opens up a new area of empirical inquiry. The hypothesis could be tested in many ways. For example, a comparison study of science students who worked with many theories as working hypotheses in a given domain and science students who did not might be attempted with respect to their sensitivity to anomalies and negative evidence.

SCIENCE TEACHERS

We have argued above that different observational languages can be used in scientific inquiry. Which observational language the scientist should use will depend on the context and purpose of inquiry. Different observational languages can also be used in the context of teaching science. Which observational language the teacher uses will depend on the context and purposes of the teaching situation. Let us briefly consider two of these educational contexts.

First, the appropriate observational language used by the teacher will depend on the level of his pupils. Some good teachers seem to sense this level instinctively, and adjust their observational language accordingly. As the class progresses and gains theoretical sophistication, the observational language of the teacher changes too. What at the beginning of a class was described as a long white line may be described later as a track left by electrons and still later as condensed droplets of water in ionized gas molecules; what was originally called a darkening of the sun may intermittently be called a shadow passing over the sun and finally an eclipse of the sun.

The general idea here as in other pedagogical contexts is to first use concepts that are familiar to the student, and to introduce new concepts and descriptions as he gains knowledge and sophistication. Failure to heed this principle in the use of observational languages may not only lose the student's interest but actually prevent him from observing phenomena in the relevant way. The naive student cannot *observe* that there is an eclipse of the sun, and it would be absurd to demand that he should.

Secondly, a good teacher will shift to a different observational language in order to bring out some contrasts and differences between theories. As we have seen, rival theories in a given domain are sometimes not readily comparable because they use different observational languages. Students may not see the theories as rivals unless their seemingly incomparable consequences are translated into observational languages that show a clear conflict. The good teacher will be able to say in a language his students understand that theory T_1 says that such and such will happen while theory T_2 says that it will not.

However, just because a good teacher often instinctively adjusts his observational language to the level of his students and the problem at issue does not mean that this tendency cannot be developed in a teacher. What the best way is to develop this tendency is unclear at the present time. But it seems likely that at least practice in observational language adjustments in different educational contexts is essential. Furthermore, knowledge of how children see the world would certainly be relevant. Teacher training in science education has unfortunately not usually stressed the importance of learning how young science students observe. Certainly more courses in genetic psychology would help, but such study by itself would hardly be sufficient. A purely intellectual knowledge of how young science students observe would be useless unless teachers were sensitive to the differences between adults' and children's observations.

One suggestion that might be helpful for developing this sensitivity would be for student science teachers to become observers in science classrooms, not from the point of view of the teacher, as is usually the case, but from the point of view of the student. The student science teacher might profit in using a technique common in the social sciences: participant observation. He might become a participant observer in a science classroom in order to learn to become sensitive to the differences between adults' and children's observations. In the next section we will consider the relevance of participant observation in social science and consider in more detail its use in education.

PARTICIPANT OBSERVATION

Participant observation is a method of observation used in the social sciences, especially in cultural and social anthropology, in order to gather information and test hypotheses.

The term "participant" in participant observation means different things.[25] Sometimes a participant observer of a community or group is a person who lives in the community or group, adopting some of their customs and habits for the purpose of observing the community or group. At other times more is involved. Some participant observers not only live in a community or group, but are in some degree assimilated into the community or group. At still other times a participant observer is a person who takes on the role of an observer of the community or group, although he may not be living in the group or may not be assimilated into it.

Whatever the meaning of "participant" in participant observation, the important question here is that of observation. What advantages for observing are there in being a participant? First, being a participant may put the observer in a position to visually attend to (noncognitively observe) certain properties he is interested in, e.g., the economic development of a community or its religious ceremonies. Without participating the observer would have to learn about such properties indirectly, such as by the use of informants. Secondly, being a participant may enable the observer to gain certain background information that allows him to observe what he would otherwise be unable to observe. Thus, living in a community may enable an anthropologist to cognitively observe what the natives cognitively observe, e.g., that Abuta is performing the ascension ceremony. This would not mean, of course, that he could not learn about Abuta and his performance without cognitive observation. A good informant might be able to give him all the information he needs about Abuta and his performance.

However, being a participant may also have certain disadvantages with respect to observation. For example, being assimilated into a community may cause the anthropologist to identify too much with the natives. Such identification may in turn cause him to acquire misinformation that results in incorrect observational judgments. An extreme case of this would be an anthropologist who went completely native and came to believe some of the superstitions of the natives. These beliefs might influence him to report, "I observe that Abuta is driving out the evil spirits." Such an observation report would be mistaken because Abuta is not driving out the evil spirits, although he might think he is.

Good anthropologists have learned to guard against this kind of problem and remain to a certain extent detached from the native mentality. Indeed, the trained anthropologist usually goes to a community with knowledge the natives do not possess, which enables him to cognitively observe what the natives could not observe. Thus an anthropologist with a knowledge of functional analysis may be able to indirectly cognitively observe that Abuta has as one of his functions preserving tribal harmony by attending to his performance of certain ceremonies. The natives, never having heard of functional analysis, cannot cognitively observe what the anthropologist does, although they are attending to, i.e., they are noncognitively observing, the same thing. Thus the warning given to anthropologists not to go completely native may not only prevent them from making mistakes in their observational reports, but enable them to observe what the natives cannot observe.

There is no doubt that participant observation is sometimes needed or desirable in the social sciences to test hypotheses and gather information. The information needed to test certain hypotheses might be obtainable only by participant observation. In certain societies reliable informants and written documents might not be available and participant observation might indeed be needed. Furthermore, there might be times when, although participant observation is not absolutely essential, it is easier to use than other methods.

In anthropology at least, participant observation is used widely. One might well wonder if this widespread use is really justified. Other methods are often available and are much easier to use. The use of informants, survey methods, written documents, and archeological reconstruction are all methods of gathering information about groups or cultures from a distance. Furthermore, nonparticipant observation methods may in certain cases enable a scientist to get the information he needs, although such methods might prevent him from making the observations that a participant observer can make. A nonparticipant observer may be able to observe that a man dressed in animal skins is dancing around a fire, but he might not be able to observe that Abuta is performing the ascension ceremony as a participant observer could. Nevertheless, the nonparticipant observer might be able to infer from other things he has observed or heard that the man dancing around the fire is Abuta performing the ascension ceremony. There is no particular value in the participant observer's observation over the nonparticipant observer's, so long as both arrive at the same conclusion. And there may be no reason why they could not.

It is also important to realize that participant observation might have great importance for achieving other goals without having any particular relevance methodologically. Thus participant observation might give anthropologists great pleasure, enrich their lives, create an esprit de corps among them, and so forth. All these things might be true *without* participant observation being a good way of obtaining information about a culture or community.

THE EDUCATIONAL VALUE OF PARTICIPANT OBSERVATION

One value that participant observation might have, even if it does not have the methodological value that is usually claimed for it, is educational. For instance, living in an alien culture and adopting some customs and habits of the natives may be an enriching and humanizing experience. Whether participant observation is the best way to achieve this enrichment and humanization is, of course, another question; whether participant observation will in all cases have these results is even debatable. Nevertheless, participant observation as a way of achieving attitude change may be an educational method well worth further exploration. One could imagine an experiment in which a group of children became participant observers of minority groups in their city. The purpose of the participant observation would not be to enable the children to gain certain information about the groups that perhaps could be better achieved in other ways. The purpose (which might be hidden from the children) would be to see if their experience had a liberalizing effect on their attitude towards these groups. Change of attitude could be tested against some control groups in which participant observation was not used.

But what relevance does participant observation have for science education? We have already suggested that student science teachers might profit from being participant observers in science classes. They might not get any more information about genetic psychology in this way than by reading a textbook. But participant observation might have other advantages: it might sensitize them to certain things that reading a textbook could never do, in particular, to differences between adults' and children's cognitive observations.

Furthermore, participant observation might have relevance not only to student science teachers, but to science students as well. Scientists themselves may be thought of as forming a particular social group, with its own values, attitudes, and customs. Students of science might well become participant observers of this group; they might live for extended periods of time among scientists, adopting their customs and habits. Unlike participant observers in anthropology, students attached to scientific groups would not have as their primary aim gathering certain vital information about the group, but rather actually learning the practice of science, achieving the scientific know-how and habit of mind characteristic of scientists.

Nevertheless, one aspect of the participant observer's role should not be given up. Just as the anthropologist does not allow himself to go completely native for fear of losing perspective and detachment, so the science student as a participant observer in a scientific community should not allow himself to be completely swallowed up in the practice of science. He ought to remain somewhat detached and critical. Such an attitude would be especially appropriate for students of science who did not plan to make science their career. For the liberal arts student in college or the typical science student in high school with no plan for a science career, participant observation in a scientific community might convey the spirit of science, instill the attitudes characteristic of it in a powerful way, and yet in a way that was not uncritical or conducive to mindless involvement.

Whether this somewhat detached attitude of the participant observer is appropriate for the science major bent on a scientific career is another question. It is not necessarily to be discouraged. Critical reflection on and passionless description of science is logically compatible with good scientific work, and under certain circumstances may even complement it.

Just as the visiting anthropologist comes to a society armed with information and skills he has acquired elsewhere that help him to understand and get along in the society, so the visiting science student in a society of scientists can come armed with information and skills he has acquired elsewhere that help him learn the practices of science and at the same time maintain a critical and detached attitude toward science. Some of this information and these skills may be acquired in a philosophically oriented course in science. Such information and skills may enable a student to observe what practicing scientists without this information and skills cannot observe. A student of scientific methodology living among scientists might cognitively observe that Dr. Smith has a background and training that enable him to observe what the student cannot observe, or to observe that Dr. Jones has accepted a hypothesis that is methodologically suspect. Thus the study and practice of philosophy of science may enable one to observe what the average scientist cannot observe, just as a scientist can observe what ordinary men cannot observe.

•

CHAPTER FIVE

The Goals of Science Education

•

So far in this book we have analyzed some fundamental concepts of science: scientific inquiry, explanation, definition, and observation. In the course of our discussion we have related our analyses to certain important issues in science education. Implicit in our discussion of science education has been the question of what science education should aim at, what its goals should be.

We have seen how science educators might become interested in the goals of science education. Science educators like Mr. Goodwin are concerned with how the goals of science education should be stated. Mr. Goodwin has become convinced that the behavioral goals approach is a good method of formulating goals. On the other hand, science educators like Mr. Brown are concerned with the normative issue of what the goals of science education should be and how these goals can be rationally justified. Mr. Brown is worried that perhaps all goals of science education are arbitrary.

The time has come to consider the goals of science education explicitly. In this chapter we will analyze some of the goals of science education, and in so doing point out their limitations. We will critically consider some of the common goals of science education, suggesting improvements when this is possible. Furthermore, we will consider how the goals of science education might be validated or tested. Toward the end of the chapter we will consider how the goals of science education make it necessary to conceive of science education more broadly than has been traditional.

FOUR ASPECTS OF THE GOALS OF SCIENCE EDUCATION

KNOWLEDGE

One of the traditional goals of science education has been the acquisition of knowledge. The kind of knowledge that is supposed to be acquired is propositional knowledge, knowledge that something is the case. Such knowledge could be of specific facts, of scientific terminology, of scientific conventions, of the structure of scientific theories, of the classifications and categories used in science, of the criteria for the evaluation of scientific theories, of methods of inquiry, and so on.

Propositional knowledge is usually represented schematically as follows:

(1) X knows that p,

where X is a person or group of people and p is some proposition. Some examples would be:

(a) Jones knows that Redi believed that maggots were not spontaneously generated in meat.
(b) Smith knows that sodium salt burns yellow.
(c) Evans knows that two of the basic concepts of biology are cell and organism.
(d) Peters knows that $T^2 = k \, \bar{R}^3$.
(e) Brown knows that a major cause of lung cancer is cigarette smoking.
(f) McGuire knows that the metal expanded because it was heated.

It is important to observe that some statements containing the word "know" do not have the same form as (1). Nevertheless, these statements are roughly equivalent to statements having the form of (1). Consider, for example:

(g) Smith knows what color sodium salt burns.
(h) McGuire knows why the metal expanded.

Now (g) is equivalent to (b), and (h) is equivalent to (f).
Sometimes, however, an equivalence cannot be constructed.

Consider:

>   (i) Miller knows how to set up the experiment.

(i) is ambiguous. It could mean:

>   (j) Miller knows that the experiment is set up in such and such a way.

In this case (i) is analyzable in terms of propositional knowledge. However, (i) could also mean:

>   (k) Miller has the skill to set up the experiment.

In this case (i) is not analyzable in terms of propositional knowledge, since Miller might have the requisite propositional knowledge and yet not have the skill. As we shall see, this sense of knowledge—which we might call the *skill sense of knowledge*—is important to science educators.[1] Let us return for the moment, however, to propositional knowledge.

   Three necessary conditions for (1) are:

>   (2) p is true.
>   (3) X is justified in believing that p.
>   (4) X believes that p[2].

   Let us consider briefly the rationale for considering (2), (3), and (4) as necessary conditions for (1). It seems quite clear that (2) is a necessary condition for (1), for unless it is, some propositional knowledge could be false. But surely this is unacceptable. It is mistaken to say, for example, that Smith knows that the earth is flat. Smith might *think* that he knows that the earth is flat. However, he cannot know that it is flat if it is not flat. The importance of the truth requirement on propositional knowledge is shown in judgments made in retrospect about whether someone knew that p. For example, today one might judge that Smith knows that p. But next year, in the light of new evidence, it might become clear that p is false. In this case the natural and correct response would not be to say that Smith knew that p, despite p being false, but rather that one had wrongly supposed that Smith knew that p.

   That (3) is also a necessary condition for (1) seems clear. Surely one wants to rule out as knowledge cases in which a person has a true belief, but a belief that has no backing or support. Consider, for example, a situation in which Brown bases his belief that a major cause of lung cancer is cigarette smoking on a dream he has. He has read no study, seen no evidence, heard no authorities, and so on; the sole basis of his belief is his dream. Although his belief is true, one surely would not say in this case that he *knows* that a major cause of lung cancer is cigarette smoking, although we would be willing to say that he *believes* this.

(4) also seems to be a necessary condition for saying that X knows that *p*. This is seen most clearly in first-person cases. It is surely absurd to say, "I know that sodium salt burns yellow, but I don't believe that it does." There are some third-person cases, however, which seem to cast doubt on (4) as a necessary condition of propositional knowledge. One might say, "Jones knows that his father is dead but he cannot bring himself to believe it." However, this seems to involve a special sense of belief. One might not emotionally accept that one's father is dead and yet cognitively accept it. Emotional acceptance is not a necessary condition for (1), but cognitive acceptance is. If Jones did not accept his father's death in a cognitive sense, then it surely would be incorrect to say that he knew his father was dead. We shall be concerned here only with belief in the sense of cognitive acceptance.

Despite the importance of propositional knowledge for science education, the nature of such knowledge has often been misunderstood by educators. In particular, the three necessary conditions of propositional knowledge considered above have been neglected. Thus Bloom et al. in *Taxonomy of Educational Objectives Handbook I: Cognitive Domain* talk about what we have called propositional knowledge, but they seem to identify such knowledge with simple recall, i.e., "bringing to mind the appropriate material." They say,

> By knowledge, we mean that the student can give evidence that he remembers, either by recalling or by recognizing, some idea or phenomenon with which he has had experience in the educational process.[3]

To be sure, if their definition is intended as a stipulative definition, there can be no objection except that it seems unnecessary and misleading to use the word "knowledge."[4] However, it is not clear that a stipulative definition is intended.

If a reportive definition is intended, there are several basic problems with this account, for there is, as we have seen, more to propositional knowledge than bringing to mind the appropriate material. First, it is not clear that Bloom et al. recognize a truth requirement. One might be able to bring to mind certain material or information that is false. Yet this is not knowledge, for someone cannot know that *p* if *p* is false. For example, a student might recall something he learned in his geography class, that Maine is the northernmost part of the United States excluding Alaska. However, he could not be said to know this since it is not true.

Secondly, Bloom et al. do not seem to recognize the need for a justification requirement. One can recall material that one has no right to believe. For example, a student might recall that there is intelligent life on other planets from his reading of a book which he knows is unreliable and untrustworthy. But we could hardly claim that the student knows that there is intelligent life on other planets, given his unreliable source of information.

Thirdly, it is unclear whether Bloom et al. take into account a belief requirement on propositional knowledge. If $T^2 = k\,\bar{R}^3$ is brought to Peters' mind at a particular time, does this entail that he believes $T^2 = k\,\bar{R}^3$ at that time? It would not seem to, but if it does not, then this constitutes another error in Bloom's account, since Peters could not know that $T^2 = k\,\bar{R}^3$ unless he believed that $T^2 = k\,\bar{R}^3$.

Whatever the problems in Bloom et al.'s account of knowledge, a further distinction is important in any discussion of propositional knowledge and science education. There are at least two different kinds of justification for believing that p. First, there are justifications that are extrinsic to the subject matter under discussion. Thus Jones' justification for believing that sodium salt burns yellow might be that his teacher or his textbook said so. Such a justification, however, is not the sort that practicing chemists would or at least should give. Secondly, there are justifications intrinsic to the subject matter.[5] A reason for believing all sodium salt burns yellow that is intrinsic to the subject matter is that investigators have burned sodium salt and it has always burned yellow.

Now some science educators may want to argue that all propositional knowledge acquired in science classes should be based on justifications intrinsic to the subject matter. Such a view is hardly plausible, however, for to carry it out would be too difficult and time consuming. Consider, for example, knowledge about the history of science, e.g., knowledge about Redi's experiment. Although it might be pedagogically advisable in certain classes specializing in the history of science to require the students to consider the historical reasons for our beliefs about Redi, in the average science course this would surely be unwise.

A more plausible position might be to maintain that only scientific knowledge—as opposed to knowledge about science—acquired in the classroom should have a justification intrinsic to the subject matter. However, even this suggestion is debatable. Why, after all, should *all* scientific knowledge acquired in the classroom be acquired with an intrinsic justification? A good practical reason for not requiring this is that it would limit severely the amount of scientific knowledge a student could acquire in the classroom. For to learn a justification intrinsic to the subject matter often involves mastering a wide range of data and its interpretation, and this takes a great deal of time. Furthermore, such time is often wasted, given the purpose of the course.

For example, a teacher may require a student to know that the distance of Mars from the sun is 1.52 times the distance of the earth from the sun. Given this information, the student may be required to compute the period of Mars by means of Kepler's third law. If his computations are correct, the student would know that Mars' period is 1.88 earth years with a justification intrinsic to the subject matter. But he may not know the distance of Mars to the sun in this way. His justification here would be based on the authority of his teacher or his textbook. However, given the purposes of the teacher, e.g., to have the student learn to use Kepler's laws and understand the function of laws in science, there would be no need for the student to know Mars' distance from the sun with a justification intrinsic to the subject matter. Indeed, such a justification would be pedagogically irrelevant and perhaps even distracting.

This does not mean, of course, that no justifications intrinsic to the subject matter should be acquired. When propositional knowledge should be acquired on the basis of justification intrinsic to the subject matter and when it should not depends on the level of the course, the topics under consideration, and other factors.

SKILLS

As we have seen, knowledge in the skill sense is not analyzable in terms of knowledge that something is the case. Having students acquire this sort of knowledge has been one of the goals of science education. Science educators have suggested a variety of skills for science students to master, depending on their level and background: specific skills such as dissecting small animals and general skills such as critical thinking; relatively pure physical skills such as setting up simple laboratory equipment and relatively pure mental skills such as deducing consequences from hypotheses.

Skills are not inborn but are mastered through practice. Thus, education in certain scientific skills is different from the education involved in acquiring propositional knowledge. One does not need to practice in order to know that sodium salt burns yellow, but one does have to practice in order to know how to set up an experiment.

It is useful to distinguish two types of skills found in science that science educators might want their students to master.[6] First, there are skills that require attention and thought to perform. For example, critically evaluating scientific papers in a certain domain of research takes a skill that is not reducible to routine; it requires careful thinking and keen attention. Let us call these skills *critical skills*. Secondly, there are skills that are capable of being reduced to routine, skills that can become automatic, as in skill in setting up a well-known experiment or in using a certain notation. Let us call these skills *facilities*. It should be noted that many critical skills presuppose facilities. Thus in order to have the skill of critically evaluating scientific papers in a certain domain, it may be necessary to have a number of facilities, for example, the facility of reading the notation in the papers.

Two cautions should be noted, however. First, to say that certain skills are reducible to routine, that they are performed automatically once they are mastered, does not mean that they are learned in a routine or automatic way. Indeed, in most cases they are not. Skill in setting up a simple experiment, for example, may be routine and automatic for mature science students, while for beginners it may take great concentration and thought. Secondly, to say that critical skills require thought and concentration does not mean that there cannot be levels of performance in the exercise of these skills. One person may critically evaluate the scientific literature in a certain domain competently, another person may do it very well, and a third may do it brilliantly.

The second point is important for science education. Science educators who have as one of their goals having their students acquire certain critical skills need not require that these skills be mastered at the highest level of proficiency. Indeed, it would be unrealistic and impractical for science educators to aim for this. In most cases, the best they can hope for is that students learn to appreciate excellent performance in a scientific skill, and that certain students are stimulated to try for the highest level of performance.

UNDERSTANDING

Another traditional goal of science education has been to get students to understand physical phenomena, theories and laws, methods, and even science itself. Very often it is not clear what is involved in this understanding.[7] However, in most cases understanding seems to be reducible to propositional knowledge or skills or both.

In some contexts understanding consists of having certain kinds of propositional knowledge. Thus to say:

(1) X understands Y

where X is a person or group of persons and Y is a thing or phenomenon is often equivalent to:

(2) X knows that $P_1, P_2, \ldots P_n$

where $P_1, P_2, \ldots P_n$ are certain propositions about Y.

Consider, for example, the following expression:

Smith understands the recent eclipse of the sun.

In many contexts this expression is equivalent to an expression about propositional knowledge. In particular, in scientific contexts it would normally be equivalent to an expression about propositional knowledge of nomological relations. Thus to say that Jones understands the recent eclipse of the sun would in most cases be to say that Jones knows the nomological connections between the eclipse and earlier events, later events, and events occurring at the same time. The extent and depth of understanding would be a function of the number of nomological connections that were known and the detail and clarity with which they were known.

In other contexts understanding involves more than propositional knowledge. It also involves having certain mental abilities and skills with respect to the knowledge.

Consider the following expressions:

Jones understands Kepler's laws.
Smith understands Poincaré's argument for conventionalism in geometry.

To say that Jones understands Kepler's laws would be to say not only that he has certain propositional knowledge — for example, knowledge of what the laws are, that they developed in certain ways, that they have such and such an application — but also that he has the ability to apply the laws, to translate them into other terms, to derive a close approximation of the laws from Newton's theory, to analyze the concepts in the law. To say that Smith understands Poincaré's argument would be to say that he has certain propositional knowledge — for example, knowledge of what the argument is, why he gave it, how others have used it or misused it, and also that he has the ability to apply the argument to new cases, to reformulate it to meet objections, to defend it against invalid criticism, and to evaluate it.

Thus in many contexts (1) is equivalent not to (2), but to:

(3) X knows that $P_1, P_2, \ldots P_n$, and X has skills $S_1, S_2, \ldots S_n$

where $S_1, S_2, \ldots S_n$ are skills connected with Y. What this knowledge and these skills consist of will depend on the context.

The problem of whether (1) is equivalent to (2) or to (3) is relevant to what people mean by understanding science. In some contexts, to say that a person understands science is simply to say that he has certain propositional knowledge about science. Such knowledge would not, of course, be of the detailed or particular findings of particular sciences, but would be of the more general or abstract properties of science: its methodology, its criteria of acceptance or rejection. (In view of this, it is hardly surprising that some educators have argued that the study of philosophy of science is important in understanding science.)

However, sometimes something more is meant by understanding science. In this sense to say that someone understands science is to say not only that he has certain propositional knowledge, but that he has certain abilities: for example, he has skill in drawing out the implications of scientific research and findings for religion and morality, he has skill in seeing connections between realms of scientific knowledge, he has skill in analyzing scientific concepts and procedures, and so on.

PROPENSITIES

An important type of goal for science educators is developing propensities in students. Propensities are tendencies to behave in a certain way. To say that someone has a propensity to do X is to say that he tends to do X, i.e., doing X is part of his characteristic behavior. In this respect skills and propensities are quite different. Having a skill in evaluating arguments is perfectly compatible with seldom exercising this skill. Thus it is consistent to say that Jones has the skill of evaluating arguments but seldom does so when he is given the opportunity. On the other hand, to say that someone has a propensity to evaluate arguments is to say that he always, or at least very often, does evaluate arguments when given the opportunity. Propensities are important for the following reasons. It is generally agreed that science should make a difference in how people conduct their lives. However, having propositional knowledge, skill, and understanding does not necessarily mean that one will conduct one's life in a particular way. A medical doctor with great knowledge, skill, and understanding of health and medical matters may not apply this knowledge, skill, and understanding to his own family, while a brilliant professor of logic and scientific method may rely on pseudoscientific theories and fallacious arguments in buying his groceries. Thus it is plausible to argue that science educators should aim not only at having students acquire knowledge, skill, and understanding, but also at having students acquire the propensity to use such knowledge, skill, and understanding in their lives.

To say that one has a propensity to act in a certain way does not necessarily mean that this propensity is a habit. For "habit" suggests some unthinking, automatic response. A propensity no less than a skill can be critical. One propensity that science educators might want students to acquire is the propensity to read certain literature critically. This is a propensity to use a critical skill, a skill that cannot be used automatically and without thought. Another propensity that science educators might want students to develop is the propensity to tidy up the laboratory when they are finished. This propensity can be unthinking and can become automatic, i.e., it can become a habit. Like skills, critical propensities often presuppose noncritical ones. Thus the habit of using a dictionary to look up new words may be essential to the critical propensity of reading critically.

In the next section we will consider some of the propensities that have been suggested as goals of science education.

CRITICAL EVALUATION OF SOME GOALS OF SCIENCE EDUCATION

ACQUIRING SCIENTIFIC PROPENSITIES

One of the most eloquent statements of the goals of science education is the statement of the Educational Policies Commission in *Education and the Spirit of Science.*[8] The Commission specified seven values which underlie science and which, in its opinion, should be acquired in education. The seven values are these:

    (1) Longing to know and understand;
    (2) Questioning of all things;
    (3) Search for data and their meaning;
    (4) Demand for verification;
    (5) Respect for logic;
    (6) Consideration of premises;
    (7) Consideration of consequences.

The Commission can be understood as specifying, by means of these seven values, seven propensities that science students should acquire in their education. Consider value (2), "Questioning of all things." One can plausibly interpret this to mean that science education should aim at students' acquiring the propensity to question all things. All the other values seem easily interpreted in this way also, that is, as propensities to be acquired.

The goals of science education suggested in *Education and the Spirit of Science* are certainly admirable ones and, as we shall see, they are a great improvement over other goals that are assumed or utilized in the field. However, these goals can be improved on and certain slight qualifications must be made before they are completely acceptable.

First, propensity (4) should be changed from "Demand for verification" to "Demand for empirical test." Propensity (4) as it stands presupposes some sort of confirmation approach to theory testing, but this approach need not be presupposed. "Demand for empirical test" would be neutral between the confirmation and refutation approaches discussed in Chapter One, and this would be all to the good.

Secondly, in elaborating on propensity (3), "Search for data and their meaning," the Commission says,

> The longing to know is the motivation for learning; data and generalization are the forms which knowledge takes. Generalizations are induced from discrete bits of information gathered through observation conducted as accurately as the circumstances permit.[9]

This passage may suggest to the reader a false picture of science. It suggests that scientific knowledge is obtained only by inductive generalization. It also suggests that scientific hypotheses are always generated by induction. As we have seen, these ideas are mistaken. Some scientific theories are not inductive generalizations and some scientific hypotheses are not generated by induction. Moreover, the phrase "discrete bits of information" may suggest that data are not conceptualized in terms of some theoretical framework. But as we have seen, this is also mistaken; there is a close connection between theory and data; data are often seen not as discrete and isolated but as unified by a theory.

The Commission seems, somewhat inconsistently, to recognize the importance of theory and of procedures which are not inductive in their other statements. In the second paragraph of their explanation of "Search for data and meaning," they say that our understanding of the world is found not in data, but in theories — *conceptual schemes*. "The evolution of those conceptual schemes is an intuitive, lengthy creative process. It involves seeing connections and meanings others have not seen."[10] However, it is difficult to reconcile this passage with the passage quoted above. Theories, the Commission seems to suggest, are also forms of scientific knowledge, they are not always generalized by induction, and they permit one to see data in a nondiscrete way.

In any case, the propensity to search for data and meaning, understood correctly, does seem to be a worthy goal of science education. This propensity should be understood in such a way that it does not presuppose any particular way in which hypotheses are generated or any objectionable theory of observation. "Search for data and meaning" can, then, best be understood as a propensity to "search for plausible and testable hypotheses that make data meaningful (understandable)." So construed, this is a propensity in the context of generation. On the other hand, the propensity to demand an empirical test is a propensity in the context of testing.

Thirdly, propensities (6) and (7) are rather curiously separated by the Commission. The reader of *The Spirit of Science* may get the idea that the propensity to consider premises is one thing and the propensity to consider consequences is another. However, in the actual context of testing, as we have already seen, these two processes go hand in hand. One tests one's premises by testing their logical consequences. It is, in general, impossible to examine the premises independently of the consequences of the premises. Thus it would be desirable if propensities (6) and (7) could be stated as one propensity—the propensity to examine premises by examining the truth or falsity of the logical consequences of these premises.

Fourthly, there is little explicit discussion in *The Spirit of Science* of the motivation for these propensities. It is conceivable that a student could have a propensity to question all things because he had the propensity to make a lot of money and he believed that questioning all things would make him a lot of money. Yet a monetary motivation would not be in keeping with *The Spirit of Science*. The closest the Commission comes to discussing the motivation for the propensities it recommends is in the discussion of propensity (1), the longing for knowledge and understanding. Here it is suggested that the spirit of science seeks to understand because "it accepts knowledge as desirable in itself."[11]

This leaves open the question of whether students should question all things because it is desirable in itself or because it is a means of achieving knowledge, which is desirable in itself. In any case, to expect beginning science students to seek knowledge because it is desirable in itself or to question all things because this is a means of achieving knowledge is asking a great deal. Surely other more realistic and less abstract goals are open. David Hawkins, for example, has suggested that "aesthetic motivation, the absorption in subject matter for its own sake," is essential in science education.[12] Students who have this aesthetic motivation, however, need not believe that knowledge is desirable in itself. Science for these students is just fun and joy.

There is no doubt that this aesthetic motivaton may be extremely important, especially in young children. The elementary science teacher who does not utilize the intrinsic joy of scientific investigation is certainly missing a good bet. On the other hand, aesthetic motivation may not be essential in all cases and other motivations may play the most important role for certain individuals at certain stages of their scientific development. At the beginning of their scientific education, at least, students should be motivated by whatever works best, whatever creates the desired propensities. Their motivation should become purified if necessary, or more sophisticated, as their education progresses. Desire for knowledge for its own sake is the sort of motivation to aim for and not necessarily the sort of motivation to demand at the start.

Fifthly, the emphasis on acquiring certain propensities should not make us overlook the fact that acquiring certain knowledge and skill is also important. Presumably one could have a propensity to seek knowledge and understanding without any skill in acquiring them; but without these skills the propensity would be useless. Again, one might have the propensity to question all things without having any skill in doing so; unless one had the necessary skills, e.g., in logic, such a propensity would be pointless. Thus it is important to understand the propensities suggested in *The Spirit of Science* and modified above as presupposing certain skills—both critical skills and facilities.

Furthermore, science students should acquire not only propensities to use these skills, but also certain propositional knowledge about the world, about scientific method, and about history. Thus a student with a great drive to utilize his skills in hypothesis generation, testing, and evaluation, critical thinking, and so on, would surely not be considered well-educated in science until he had some basic knowledge of the world he lived in and of the history of science.

Thus, the emphasis on propensities should not blind us to other aspects of a complete scientific education. *The Spirit of Science*, although a welcome antidote to the heavy stress on acquiring propositional knowledge that is so characteristic of much of traditional science education, perhaps neglects one good aspect of traditional science education: the acquisition of propositional knowledge. Nevertheless, despite the limitations of *The Spirit of Science* and the qualifications that have to be made with respect to it, the goals specified there do seem superior to many goals of science education found explicitly or implicitly in the literature. Let us consider some of the alternatives.

LEARNING TO PREDICT

One common suggestion about what science education should aim at seems to be based on a mistaken view about the nature of science. It has been argued that the principal purpose of science is to gain concepts by which the behavior of things can be predicted. Consequently, it has been assumed that the principal purpose of science education is to have students of science acquire these predictive concepts.[13]

There are a number of mistakes in this suggestion. First, science does not predict by means of concepts. It predicts by means of theories, hypotheses, and laws. The concepts of mass and energy, for example, do not predict anything. To be sure, theories, hypotheses, and laws may contain certain concepts, but the concepts taken in isolation have no predictive power. Secondly, even if a student acquired some theory or hypothesis that enabled him to predict, this would still not mean that he had any desire to use this theory or hypothesis, or any ability to judge its predictive capacity. A much more plausible goal of science education is to have students acquire this desire and ability to predict.

The above problem can be overcome by modifying the thesis in the following way: The principal goal of science is to acquire systems—theories, hypotheses, and so—by means of which the behavior of things can be predicted. The principal goal of science education is to foster in science students the desire to use these predictive systems and the ability to make judgments about the predictive powers of these systems.

There is no doubt that this construal is an improvement over the first, but it still seems to have certain problems.

First, as we have seen in Chapter Two, it is a mistake to think that the major goal of science is prediction or even explanation. The major goal of science is achieving scientific understanding of the world. This involves seeing phenomena in their nomological connections with each other. Knowledge of some—but only some—of these connections enables us to predict. Prediction may be the principal goal of applied science, but in theoretical science it plays a more subordinate role.

Secondly, it is not enough to say that students should have the desire to use predictive systems and the ability to judge whether a system is predictive. Students might have this desire and yet do nothing about it. Surely it is more plausible to suppose that students should have the propensity to use predictive systems and not just the desire. Again, students might have the ability to judge whether a system was predictive and yet not judge whether any are. What is wanted is not just the ability, but the propensity. But even this more plausible suggestion must be qualified in certain ways. Science students might acquire the propensity to use unscientific systems. Suppose one could use astrology to make good predictions. One might suppose something had gone seriously wrong in science education if science students were taught to use such systems in making predictions. What is surely wanted is rather the propensity to use *scientifically acceptable* predictive systems. However, as we have seen, we surely do not want students merely to acquire this propensity. We want them to have a more inclusive propensity, a propensity to seek scientific understanding and use this understanding in their lives. The propensity to use predictive systems is only part of this propensity.

One commonly stated goal of science education is the goal of understanding science.[14] As we have seen, understanding science can mean two different things: merely having knowledge about science — in particular, knowledge about its general history, method, and problems; or also having certain skills and abilities.

Understanding science in this latter sense should undoubtedly be one of the major goals of science education — at least at certain levels — and recent emphasis on this goal has been salutary. However, an exclusive emphasis on this goal is too limited. First, understanding science in both senses apparently can be achieved with only the barest understanding of natural phenomena. One might understand science but not the fall of a body, the burning of paper, or the behavior of human beings. But surely one wants students to understand natural phenomena as well as the enterprise that studies natural phenomena.

Secondly, understanding in both senses is completely cognitive; it implies no attitude or orientation toward life. One might understand science — and even understand the phenomena of the world — and yet have a negative attitude toward science, have no propensity to approach life and its problems scientifically, and be the most unscientific of thinkers and actors. But surely what is wanted is for students not only to understand science, but to have a scientific attitude and approach toward the world and its problems. To put it another way, what is wanted is that students have a propensity to use their scientific understanding.

Another goal of science education — usually implicitly assumed rather than stated — is to get students to think like scientists, to see the world like scientists, and so on. There would be nothing objectionable about this goal if one had good reason to suppose that scientists were perfect specimens both cognitively and otherwise. Unfortunately, this is often not true. Often a scientist, although he may have a deep understanding of certain phenomena in a certain domain and even the correct scientific attitude in this domain, is not so admirable outside this domain. He may have no understanding of science and, moreover, he may approach life and its problems in an unscientific manner. Surely having science students emulate this sort of scientist would be objectionable.

On the other hand, one might argue that students should identify with model scientists and not just any scientists. But who are model scientists and what characteristics do they have that make them model? Furthermore, which of these characteristics should students emulate? The propensities considered above — with the suggested modifications — seem to include most of the important characteristics of model scientists. At least they seem to include all the ones worth emulation by students. So the suggestion that students should identify with model scientists seems very close indeed to the suggestion that students should acquire the propensities considered above.

It seems, therefore, that the propensities suggested by the Educational Policy Commission in *The Spirit of Science*, with the slight modifications and qualifications given them above, are an improvement over the other suggestions we have considered, or else the other suggestions reduce to the suggestions of the Educational Policy Commission. Moreover, these propensities do seem to be rationally justified as goals of science education.

TESTING THE GOALS OF SCIENCE EDUCATION

In any discussion of testing the goals of science education, two questions need to be separated. First, given that certain goals should be the goals of science education, how can one tell whether these goals have been achieved or not? Let us call this *achievement testing*. Secondly, how can one tell whether certain goals should be the goals of science education, i.e., should be accepted? Let us call this *acceptance testing*. Science education has been mainly concerned with the achievement testing of goals, and this is not surprising. Achievement testing is very practical and is straightforwardly scientific. Acceptance testing is more philosophical and normative: it does not seem to be straightforwardly scientific at all. Nevertheless, both kinds of testing are crucial for science education, since an acceptable approach by educators to achievement testing would be of little point without an acceptable approach to acceptance testing.

ACHIEVEMENT TESTING

We will consider achievement testing by critically discussing the recent behavioral goals movement in science education.

Behavioral Objectives. There is today a great emphasis on behavioral goals in science education. Indeed, as one critic has suggested, "Anyone who confesses to reservations about the use of behaviorally stated objectives for curriculum planning runs the risk of being labeled as the type of individual who would attack the virtues of motherhood."[15] However, despite this great emphasis on behavioral goals, the whole issue needs reexamination.

First, it is not clear what is meant by "behavioral goals." Several different interpretations have been suggested by recent discussions.

One gathers that by "behavioral goals" is sometimes meant what can be observed. Thus, to say that the goals of science education should be limited to behavioral goals is simply to say that the goals of science should be limited to what can be observed in students. But as we have seen from our discussions in Chapter Four, what can be observed is a function of many factors, including the background and training of the observer. A scientist can observe things because of his training and background that a layman because of his background cannot. The same is true of an experienced teacher. Thus a master teacher with the relevant background and training can observe that Johnny does not know the answer but has only guessed it, while an apprentice teacher cannot. To be sure, the apprentice may tentatively infer that Johnny does not know on the basis of his silly grin, his hesitation, and his fidgeting. What is gained through a deliberate inference by an apprentice is obvious at a glance to a master teacher.

According to this observational interpretation of behavioral goals, what is a behavioral goal is relative to the background and training of the observer. Hence, the idea of behavioral goals is a relative notion, in that what is a behavioral goal for one observer is not for another. However, this interpretation of behavioral goals is in conflict with one of the things advocates of the behavioral goal approach seem to have in mind: that behavioral goals are absolute rather than relative to the background and training of the observer.

Although advocates of the behavioral goal approach seem to assume that behavioral goals are absolute, the behavioral goals actually stated by some advocates are only observable to rather sophisticated observers, observers with specialized background and training. For example, French and his associates argue that one behavioral goal connected with logical thinking is recognizing the unsoundness of drawing generalizations from insufficient evidence.[16] But surely it takes a rather talented observer to observe this in a student. The observer would himself have to be able to recognize the unsoundness of the inference; he would also have to be able to see through the possible sham and pretense of the student; he would have to observe that the student really recognizes the unsoundness, rather than simply pretending to. Kurtz gives as a possible behavioral goal describing an object so clearly that another person can draw a picture of the object.[17] Again, it would take a rather sophisticated observer to observe this is a student.

Sometimes what seems to be meant by behavioral goals are goals which are stated in nonmentalistic terms. According to this interpretation, to say that all the goals of science education should be behavioral goals is to say that the statement of these goals should not contain mentalistic language like "know," "believe," "intend," "understand," and so on. This interpretation is sometimes closely connected with the first interpretation, since it has often been believed that the mental attributes of a person are unobservable.

However, this latter view, although quite common, does not seem to be true. As we have already seen, what is observable is a function of one's background and training. Sophisticated observers can observe that other people intend to do certain things or that they believe certain things. The master teacher, for example, can observe that Johnny does not know the answer and is only pretending.

In any case, the idea that mentalistic language should be excluded from the statement of the goals of science education is apparently not taken seriously by most advocates of behavioral goals. French and his associates, for example, state as one behavioral goal that a student "respect and use with understanding the scientific method for discovering solutions to problems."[18] Mentalistic language is explicitly used here. Even where mentalistic language is not explicitly used, mentalistic concepts seem to be tacitly assumed. Kurtz lists as an example of a possible behavioral goal stating an interpretation of some data.[19] Now to "state" that something is the case, in contrast to merely mouthing the words, involves believing or supposing that something is the case. Thus, if Jack states, "The interpretation of the data is X," we would normally take him to believe that the interpretation of the data was X. So it would seem that stating something, in the usual sense of the term "stating," implies some mentalistic concepts. Kurtz lists as another example of a behavioral goal calculating the slope of a curve.[20] "Calculating" as usually understood connotes more than moving a pencil or slide rule and arriving at a certain answer; it connotes some purposeful action, and in human beings at least this involves an intention or motive. These are mentalistic notions. So it seems that calculating the slope of a curve involves mentalistic notions.

A third meaning sometimes attributed to behavioral goals is merely specific goals, and often in the literature "specific" and "behavioral" are used almost interchangeably. Thus Kurtz argues that "providing understanding" and "enabling students to be critical" are unspecific and "specific behavioral objectives" need to be substituted. It is unclear what "behavioral" adds to the expression "specific objectives" in this context.

Goals in terms of understanding can be very specific. One goal of science education might be to have an understanding of X where X is as specific a description as one pleases. For example, X

might be "what happens when two two-and-one-half-gram weights are placed on one pan of the balance beam and one five-gram weight is placed on the other pan." On the other hand, behavioral goals can be quite general. One educational goal might be for students to manifest exploratory behavior or to respond in some way when spoken to.

Indeed, some of the behavioral goals specified by advocates of the behavioral goals approach could be much more specific than they are. (Whether they *should* be is another question.) Consider the behavioral goal given by French and his associates already mentioned—"Respect and use with understanding the scientific method for discovering solutions to problems." Clearly a more detailed specification could be given, as, for example, "Respect and use with understanding the scientific method as understood by Poincaré for the solution of problems in physics." Kurtz suggests that demonstrating how to test a prediction is an example of a behavioral goal. Here again a more detailed specification is possible, such as demonstrating how a prediction from Newton's theory is tested in a short period of time with a simple pendulum.

Thus there is nothing intrinsically specific about behavioral goals (unless "behavioral" and "specific" are used interchangeably), and there is nothing intrinsically unspecific about goals in terms of understanding or knowledge.

Another interpretation of behavioral goals is this. To say that the goals of science education should be behavioral is simply to say that the goals of science education should be stated as testable hypotheses, in particular, that they should be stated as testable hypotheses that science students have certain characteristics. There is no necessary connection between testable goals in this sense and observable goals or goals in terms of nonmentalistic concepts.

Let us consider observable goals first. Suppose that one goal of science education is for science students to acquire characteristic C. Consider the hypothesis that Billy, a young science student, has acquired C after his course in science. C might not be observable to a beginning science teacher. For example, suppose C is being able to describe an object clearly enough so that other people can draw the object. However, the hypothesis that Billy has C is still testable for the beginning teacher. The teacher might be able to infer that Billy has C on the basis of the reactions of the other students.

Now consider mentalistic concepts. Suppose C is a mentalistic characteristic of Billy, that Billy believes that Mars' period is 1.88 the earth's year. The hypothesis that Billy believes this is not untestable, since it might be inferable from his verbal and nonverbal responses.

There is a close connection between specificity and testability, but the connection is rather different from that which some science educators suppose. Suppose our goal is a nonspecific one, e.g., to have science students develop some general propensity, such as the propensity to be critical. Consider the hypothesis that Jack, a high school science student, is critical after his courses in science. Such a hypothesis is not too general to be testable. What needs to be specified in detail is the evidence that would be relevant for testing this hypothesis. Such evidence can take many forms, e.g., Jack challenges premises of arguments when they are presented to him, he tests the validity of arguments when he sees them, he questions the reliability of sources of information people cite, and so on. Thus the hypothesis that Jack is critical is not a specific hypothesis, but it is testable, since the evidence for whether Jack is critical is whether Jack responds in certain specific ways in certain specific situations.

Even here, however, the hypothesis that Jack is critical is not tested in isolation from other hypotheses, but only against a background of auxiliary hypotheses. Suppose Jack does not point out a flaw in an obviously invalid argument when it is presented to him. This would not necessarily count against the hypothesis that Jack is critical, for Jack may be overly sensitive and polite and believe that refuting the argument may hurt someone. His critical impulse may be in conflict with his desire not to offend. On the other hand, if Jack does point out the fallacy, this does not necessarily count for the hypothesis that he is critical unless other hypotheses are also assumed. Jack may have been forced to respond in this way by other members of the class; he may not have seen the fallacy at all, and may not even care.

These considerations point out another serious misunderstanding in recent discussions about behavioral goals. When behavioral goals are identified with testable goals and confused with specific goals, it is almost universally assumed that specific student responses count for or against some desirable or undesirable trait in the student, in isolation from other hypotheses about the student's psychological and cognitive makeup. However, this simple-minded view of hypothesis testing is no more true here than in other areas of science.

The Justification of Behavioral Goals. Why are behavioral goals considered so desirable and nonbehavioral goals so undesirable by science educators? The reasons given are largely based on mutually supporting misunderstandings about the nature of science and the confusing of the various theses analyzed above. Once these different theses are sorted out, much of the plausibility of behavioral goals vanishes.

First, many science educators seem to assume that science

deals with observables, and that unobservables are inadmissible in science. Since they identify behavioral goals and observable goals, they assume that nonbehavioral goals should be eliminated. Science, however, does deal with unobservable entities and processes. Electrons, superegos, and genes were all at one time unobservable to scientists. Moreover, science educators seem to have a much too simple-minded view of observation: namely, the view that something is observable independent of the background of the scientist. Once these misunderstandings are exposed, there is no basis, in the nature of science at least, for not having nonbehavioral, i.e., nonobservable, goals.

Secondly, mentalistic notions are assumed to be inadmissible in science, and, hence, in science education. This view seems to be based either on the idea that science must deal only with observables (the view we have considered above) and that mentalistic properties are not observable, or on the idea that hypotheses about mentalistic entities are untestable. But all of these considerations are mistaken: as we have seen, there is no reason why science needs to be restricted to observables and hence no reason why science education needs to be. Mentalistic properties are not unobservable per se and hypotheses about mentalistic entities are not necessarily untestable – any more than hypotheses about electrons or genes are. Thus there is no basis, at least in the nature of science, for restricting the goals of science education to behavioral goals.

Thirdly, it seems to be assumed that a hypothesis in science in order to be testable needs to be specific; hence, it is argued that the goals of science education need to be specific in order to be testable. However, as we have seen, this view is based on a confusion between the specificity of a hypothesis and the evidence for it. Many hypotheses in science are quite general and nonspecific, although particular and specific evidence counts for or against them. There is no reason, therefore, why the goals of science education cannot be nonspecific and yet testable. However, as we also have stressed, the testing of whether certain goals have been achieved is no simple matter. Hypotheses about whether a student does or does not have some desirable trait are not tested in isolation from other hypotheses about the student.

A Reinterpretation of Behavioral Goal Emphasis. The emphasis on behavioral goals has indeed had value, but what exactly the value is has often been missed because of the sorts of confusion and misunderstanding listed above. However, the slogan "Help stamp out nonbehavioral goals" can be given a plausible interpretation as a call for clear specifications of what evidence counts for or against some hypothesis about the presence of some desirable property or trait of a student. This call demands that educators clarify their goals and articulate them in such a way that they have testable implications. Thus the stress on behavioral goals may be interpreted as advocating the following rule:

> PR_3 Interpret or reinterpret the goals of science education in such a way that a hypothesis that a goal has been achieved, when combined with certain auxiliary hypotheses, has test implications of the following form:
>
> (1) If student P receives stimuli S, then student P will respond in manner R.

For example, educators have argued that having a good understanding of science is one goal of science education. Now PR_3 urges them to interpret this goal in such a way that the statement that the goal has been achieved by some student, when combined with certain auxiliary hypotheses, has certain test implications of a particular form. For example, consider the following test implication derived from the statement, "Bill has a good understanding of science," when combined with certain auxiliary hypotheses:

> (2) If Bill is given the questions on the Nature of Science Scale (NOSS),[21] at least 70 percent of Bill's answers will agree with the model responses (the "correct" answers).

It is clear that verification of (2) would confirm that Bill has a good understanding of science only if the auxiliary hypotheses used to derive (2) were true. One obvious auxiliary hypothesis is that the model responses on the NOSS reflect an enlightened opinion on the nature of science. There is some reason to doubt that this auxiliary hypothesis is true, since at least two of the basic assumptions of the test are dubious: first, there is no one scientific method but only scientific methods, and second, science insists on operational definition. These assumptions, apparently derived from the mistaken views of Conant and Bridgman,[22] may seriously affect the validity of the test, since they are reflected in what are considered to be correct answers to the questions on the test.

Another obvious auxiliary hypothesis that is needed for the confirmation of (2) to confirm that Bill has a good understanding of science is that Bill was not cheating in answering the questions. Clearly if Bill answered the questions by copying off his neighbor, then the inference that he had a good understanding of science would not be supported.

Is PR_3 a fruitful rule? No doubt it forces science educators to think about how to determine whether some postulated goal of science education has been achieved, and surely this is all to the good. On the other hand, this rule must be used with certain cautions and reservations. First, it is not always easy to specify test implications. Indeed, it may take a great deal of rational reconstruction of the terms in the auxiliary hypotheses and in the statement of the goal to bring it off. It is important to notice that PR_3 reads "interpret or reinterpret," and this interpretation or reinterpretation in many cases is a piecemeal task. One modifies and shapes and articulates the goals and auxiliary hypotheses in a groping, tentative manner. This slow process of validation is not unique to science education, but is typical of much of science. However, one gets the impression from people who advocate behavioral goals that, unless test implications are immediately derived, the postulated goals are suspect. This is not necessarily true.

Closely related to the first point is the fact that PR_3 should not be pursued vigorously in early attempts at curriculum development.[23] Early attempts at specifying test implications may frustrate the working out of theoretical relations between the hypothesis specifying the goals and auxiliary hypotheses, i.e., in developing an adequate theoretical basis. It may, in short, misdirect intellectual energy to what at the time are less important tasks.

ACCEPTANCE TESTING

So far our discussion of testing the goals of science education has centered on the question, "Given certain goals, how can one tell whether these goals have been achieved?" The recent confused emphasis on behavioral goals has been considered in connection with this question. The question remains, however, how we can tell which goals *should* be achieved. For many science educators such a normative question is decided by arbitrary fiat. It is supposed that there is no rational way to test whether a proposed goal of science education is acceptable or not, and hence that rational discussion of this question is impossible. Such an attitude is unfortunate but perhaps understandable, since it may be based on a too narrow view of rationality, which itself may be based on a too limited view of science. In any case, we will suggest three approaches to justifying the goals of science education. These approaches should not be thought of as mutually exclusive. Indeed, we will try to show that they are mutually supporting.

Instrumental Justification. One obvious way that any goal G_1 of science education can be justified is hypothetically or instrumentally: One argues that goal G_1 is instrumental to achieving some other goal G_2, e.g., some moral goal considered to be more basic. For example, one might argue that having students acquire the sort of propensities suggested by the Educational Policies Commission is instrumental in achieving a happy and just society. Such a proposal could be formulated in the following hypothesis:

H_1 Acquiring propensities $P_1 + P_2 \ldots P_n$ is instrumental in bringing about a happy and just society.

H_1 should then be considered a hypothesis to be evaluated by the standard methods of science. Instrumental validation is relative: one validates some goal relative to the acceptability of other goals. The problem with instrumental validation, it might be argued, is that it presupposes certain other, more basic goals which are themselves unjustified. This may be true, but in actual disagreements about the goals of science education, basic ethical goals are not always in question. More basic goals may be assumed by all parties in the dispute. The question then can be posed in terms of whether fulfilling certain goals of science education will help achieve the ethical goals agreed upon. So in this sense at least, rational discussion is possible; one can rationally evaluate whether achieving goal G in science education will help achieve a more basic goal of society which, although perhaps not itself justified, is generally agreed upon.

In this sense it certainly seems plausible that the sorts of propensities suggested by the Educational Policy Commission are instrumentally justified. In any case, hypothesis H_1, when $P_1 + P_2 \ldots P_n$ are interpreted as the propensities suggested by the Commission, seems a fruitful hypothesis to investigate. Moreover, most people seem to believe that a happy and just society is an excellent goal to aim for.

Naturalistic Justification. For many people, however, such an instrumental justification is not completely satisfactory. Perhaps another approach would strengthen the case. It has been argued by some philosophers that the only thing one could plausibly mean by saying that some goal is justified is that it is approved of by rational men, free from prejudice, who reflect deeply and coolly on the consequences of achieving this goal.

An analogy from the acceptability of observational reports may help us understand this point. As we have seen, observational reports are not indubitable. Nevertheless, they do have a *prima facie* claim to acceptability if they are made under standard conditions. The observer must not be in a poor light, suffer from hallucinations, and so on. Thus one might say:

(1) Observational report R is *prima facie* acceptable = R was truthfully made by a competent observer under standard conditions.

As we have seen, however, although *prima facie* acceptable, R may still be rejected. The scientist must carefully and rationally consider the effects of accepting R on this entire system of theories, laws, and observations. Thus one might say:

> (2) Observational report R is acceptable = R is *prima facie* acceptable and R is accepted after careful and rational consideration of the consequences of accepting R on the rest of the system.

A similar sort of argument could be given with respect to acceptable, i.e., justified, goals:

> (1') Goal G is *prima facie* acceptable = G is proposed by a man who is unbiased, unprejudiced, and well informed.

However, just like observational reports, such goals, although *prima facie* acceptable, may be ultimately rejected in the light of further consideration. Thus:

> (2') Goal G is acceptable = G is *prima facie* acceptable and G is approved after careful and rational consideration of the consequences of accepting G on the rest of the system of goals.

Definitions (1') and (2') should not be thought of as either reportive definitions or stipulative definitions.[24] Rather they should be construed as rational reconstructions. Such definitions seem to capture what many people mean by the *definiendum*, but they also purport to improve upon the ordinary meaning in the interests of clarity and testability. This approach to justification is called naturalistic because the *definiens* is couched in language that is subject to empirical test by the methods of science.

Construed in this way, the validation of any general goal of science, such as the propensities discussed at length above, would be in terms of the procedures specified in definitions (1') and (2'). For example, one would validate the goal of acquiring these propensities by seeing if one approved of them on reflection, in an unbiased mood, when one was well informed and had considered the consequences of accepting them for one's other goals. It seems that reflective and rational people do approve of these goals in the general form outlined above.

The naturalistic approach is perfectly compatible with instrumental justification. For reflective and rational people may (and probably would) approve of the propensities considered above as the major goals of science education because of their effect on securing certain general ethical goals. These ethical goals themselves may be accepted after reflection and deliberation; they need not be unjustified. However, reflective men may also approve of acquiring such propensities as desirable in themselves—not merely as instrumentally valuable. Indeed, as we shall see, such approval seems to be implicit in the institution of rational discussion.

Dialectic Justification. A common and powerful way of refuting an adversary in a dialectic exchange is to show him that a particular stand he is taking is belied by his whole approach to the problem, or that the opposite of the stand he is taking is presupposed in his manner of arguing. Thus a person who asserts, "No statement is either true or false," surely assumes that his statement is either true or false; indeed, he assumes that it is true.

A similar sort of move might be made against someone who questions whether some of the propensities considered above are worthwhile or acceptable goals. For someone who seriously questions the desirability of the spirit of science manifests that spirit: he tacitly assumes the critical posture and approach that is characteristic of science; his questioning is belied in his own behavior. In other words, a person who is willing to engage in critical rational debate commits himself to the desirability of the propensities typical of science, since critical debate about goals presupposes a respect for logic. Without this respect, meaningful debate would be impossible.

It would seem, therefore, that in broad outline some of the propensities specified above cannot be legitimately challenged in rational discussion without the challenger inconsistently presupposing in his behavior the very goals he is challenging. It should be noted in this respect that our criticisms of goals specified in *The Spirit of Science* were not fundamental. They involved rather minor points of clarification and qualification. Our whole critical approach presupposed in broad outline the values suggested by the Commission; it could not have been otherwise.[25]

SCIENCE EDUCATION BROADLY CONCEIVED

So far we have argued that certain propensities characteristic of the spirit of science should be the major goals of science education. We have argued that these propensities—properly understood and qualified—are not only an improvement over other commonly stated goals, but are justified in their own right. Indeed, we have argued that anyone who engages in meaningful rational debate about the goals presupposes that some of these propensities are desirable.

However, the propensities that we have advocated and their development in science education should not be conceived of narrowly. These propensities, correctly understood, are characteristic not just of the ideal scientist, but of the rational man. As the Educational Policy Commission suggests, "What is being advocated here is not the production of more physicists, biologists, or mathematicians, but rather the development of a person whose approach to life as a whole is that of a person who thinks—a rational person."[26] Thus the complete manifestation of the spirit of science goes beyond the confines of what is usually called science into practical, moral, and even religious contexts. The goals of science education should also be conceived of in these contexts.

An excellent physicist who is mindless and uncritical in buying his son a bike or himself a new car is deficient not just in his consumer education. There is something profoundly lacking in his science education. He would not dream of accepting a new physical theory without careful evaluation of the evidence. Yet he accepts the claims of the manufacturer without a qualm. For a well-trained scientific mind, the claims of the manufacturer ought to be hypotheses to be evaluated in the same objective way as any other hypotheses. Similarly, a good chemist who is uncritical of some simple-minded answer to a certain complex moral problem is not just lacking in his moral education, but is also deficient in his scientific education. The well-trained scientific mind would consider the alternatives and the relevant evidence in considering an answer to a problem in chemistry or morality.

The aim of science education ought to be to produce people imbued with the spirit of science who manifest that spirit in all relevant contexts. In order to bring about the manifestation of the spirit of science in typically nonscientific contexts, science education will have to be conceived of much more broadly. Instruction in science, for example, can no longer be considered the sort of activity that goes on in the typical science classroom. Consumer education and parts of moral education, to cite just two examples, should be conceived of as an essential part of science education.

In actual practice how might this generalized conception of science education work? Two different approaches might be taken. Consider, for example, consumer education as part of science education. First, the traditional science education curriculum might be harmonized with a consumer education curriculum. For example, the content and structure of a general chemistry course might be dovetailed with a home economics course that concentrated on consumer problems. The home economics course could illustrate many of the practical applications of chemistry in choosing products for general consumption. The chemistry course could give students the theoretical background and analytic tools to understand the practical applications. General chemistry and home economics might be required as a course sequence.

Secondly, the content and structure of the traditional science course might be changed, instead of being integrated with other courses. In a biology course, for example, illustrations and field trips could be partially devoted to consumer problems. Biology textbooks, instead of discussing Redi's test of spontaneous generation as an illustration of scientific method in action, could discuss the testing of the hypotheses that cigarette smoking causes lung cancer and high dosages of vitamin C prevent colds. Field trips, instead of consisting of a search for interesting wildlife in the local woods or parks, might be a search for biologically relevant health menaces in the community.

Consider moral education as essentially connected with science education. Again, science courses and courses devoted to social and moral questions, e.g., civics or social studies courses, could be harmonized. Certain facts about heredity and birth learned in a biology course could give students the information needed to discuss and consider intelligently in their civics or social studies courses social issues connected with eugenics, racism, and birth control. Of course the biology and civics or social studies courses would have to be designed carefully so that the necessary feedback and dovetailing occurred. The civics and social studies courses would be designed to show the relevance of biological theory and findings for social issues; the biology course would be designed to give students the information and theory necessary to understand the issues discussed in the civics or social studies courses. The emphasis throughout would be on the importance of scientific method and knowledge for understanding social issues.

Another possibility would be to raise certain moral issues in science courses themselves and attempt to attack them in the same way that other issues in the course are attacked — by the use of scientific method broadly conceived. Indeed, some moral issues might arise naturally out of the very process of conducting the course. One example might be the dissection of insects and other small animals that is typical in biology courses. Such dissection raises important moral issues that are seldom considered by the teacher, let alone brought to the attention of the pupils. For example, what moral right do we have to dissect these animals?[27]

An answer that is sometimes given is that such creatures do not feel pain. But this answer is surely not very satisfactory. First, how can we be sure that they do not feel pain? Secondly, even if they do not feel pain, what right have we to kill living creatures? Indeed, where does one draw the line? Most biology teachers and students would be horrified at the dissection of a small dog and yet they proceed without qualms to dissect a large beetle. Where along the continuum of living things is killing in the name of science permitted?

It is sometimes argued that the small animals and insects that are killed and dissected are pests. But this is not true in all cases. One of a horde of grasshoppers destroying a farmer's crops may be a pest, but a lone grasshopper in a city lot is surely not. In any case, the question remains, even if some animals are pests, does this give humans the right to kill them? Recall that some dogs and even people are considered pests. They are not usually dissected because of this.

All of this is not to suggest that students should not dissect insects and other small animals in their biology courses. What it does mean is that a decision to do so is a moral decision that can and should be made intelligently and rationally, a decision that should be made in the light of the alternatives and the evidence by teachers and by students.

Whatever the details of the integration of moral and scientific education, one thing is clear: the ways of science and the ways of morality are intimately connected. The intellectual virtues characteristic of science—honesty, objectivity, impartiality, and rationality—are moral virtues. Science education broadly conceived ought to foster these virtues in both scientific and moral contexts. Let us venture to hope that in the future the ways of science will become meaningful to students and become *their* ways.

NOTES

INTRODUCTION

1. See Joseph Schwab, "The Teaching of Science as Enquiry," in *The Teaching of Science* (Cambridge, Mass.: Harvard University Press, 1964), pp. 1–104.

2. For statements of the structure approach to science education see Jerome Bruner, *The Process of Education* (Cambridge, Mass.: Harvard University Press, 1961) and Joseph Schwab, "Structure of the Disciplines: Meanings and Significances," in *The Structure of Knowledge and the Curriculum*, ed. G. W. Ford and Lawrence Pugno (Chicago: Rand McNally & Co., 1964), pp. 1–30.

3. For an influential statement of this approach see Edwin Kurtz, "Help Stamp Out Non-Behavioral Objectives," in *Readings in Science Education in the Secondary School*, ed. Hans O. Andersen (New York: The Macmillan Co., 1969), pp. 142–145.

4. For a similar contrast between analytic philosophy of science and speculative philosophy of science see Israel Scheffler, *Anatomy of Inquiry* (New York: Alfred A. Knopf, Inc., 1963), Ch. 1, and May Brodbeck, "The Nature and Function of the Philosophy of Science," in *Readings in the Philosophy of Science*, ed. H. Feigl and M. Brodbeck (New York: Appleton-Century-Crofts, 1953), pp. 3–7.

5. For a discussion of the distinction between analytic and normative philosophy of education in general terms see William Frankena, *Philosophy of Education* (New York: The Macmillan Co., 1965), pp. 1–4.

6. See, for example, Barney Berlin and Alan Gaines, "Use Philosophy to Explain the Scientific Method," *The Science Teacher*, 33 (May 1966):52, and Merritt Kimball, "Understanding the Nature of Science: A Comparison of Scientists and Science Teachers," *Journal of Research in Science Teaching*, 5 (1967–1968):110–120. Kimball's study showed that philosophy majors scored higher on *The Nature of Science Scale* than science majors. However, as we shall argue in Chapter Five, certain basic assumptions of this test are in error.

7. For a review of recent literature relevant to some of the theses argued for in this book see James T. Robinson, "Philosophical and Historical Bases of Science Teaching," *Review of Educational Research*, 39 (1969):459–471.

CHAPTER ONE

1. See Biological Science Curriculum Study, *Biological Science: An Inquiry into Life* (New York: Harcourt Brace Jovanovich, Inc., 1968), pp. 23–26.

2. A similar distinction is sometimes made by the use of the terms "the context of discovery" and "the context of justification." See Hans Reichenbach, *Experience and Prediction* (Chicago: The University of Chicago Press, 1938), Ch. 1, Sec. 1.

3. This approach of teaching science as inquiry is usually associated with Joseph Schwab, "The Teaching of Science as Enquiry."

4. For a similar distinction see F. James Rutherford, "The Role of Inquiry in Science Teaching," *Journal of Research in Science Teaching*, 2 (1964):80–84.

5. See David P. Ausubel, "Some Psychological and Educational Limitations of Learning by Discovery," in *Readings in Science Education for the Secondary School*, pp. 97–114.

6. See Graham Wallas, *The Art of Thought* (New York: Harcourt Brace Jovanovich, Inc., 1926), Ch. 4. A fourth stage, verification, is omitted here.

7. Cf. Carl G. Hempel, *Philosophy of Natural Science* (Englewood Cliffs, N.J.: Prentice-Hall, Inc., 1966), p. 14.

8. Recent discussion of a logic of discovery has centered around the work of the late N. R. Hanson. See Norwood Russell Hanson, "Is There a Logic of Discovery?" in *Current Issues in the Philosophy of Science*, ed. H. Feigl and G. Maxwell (New York: Holt, Rinehart & Winston, Inc., 1961), pp. 20–35. For an analysis of Hanson which clarifies what Hanson was up to see Wesley C. Salmon, *The Foundations of Scientific Inference* (Pittsburgh: University of Pittsburgh Press, 1966), pp. 111–114. For a discussion of a logic of discovery in science education literature see G. L. Farre, "On the Problem of Scientific Discovery," *The Science Teacher*, 33 (October 1966):26–29.

9. Peter Shoresman, "A Technique to Clarify the Nature of Theories," *The Science Teacher*, 32 (May 1965):53–55.

10. Frank Micciche and Michael Keany, "Hypothesis Machine," *The Science Teacher*, 36 (April 1969):53–54.

11. Shoresman, "A Technique to Clarify the Nature of Theories."

12. Micciche and Keany, "Hypothesis Machine."

13. Shoresman, "A Technique to Clarify the Nature of Theories."

14. Micciche and Keany, "Hypothesis Machine."

15. Such an approach is very widespread. See, for example, Carl G. Hempel, *Philosophy of Natural Science;* Ernest Nagel, "Principles of the Theory of Probability," *International Encyclopedia of Unified Science*, 1, no. 6 (1939); Rudolf Carnap, *Logical Foundations of Probability* (Chicago: The University of Chicago Press, 1962); Arthur Pap, *An Introduction to the Philosophy of Science* (New York: The Free Press, 1962); and John Patrick Day, *Inductive Probability* (New York: Humanities Press, Inc., 1961).

16. See Karl Popper, *The Logic of Scientific Discovery* (New York: Basic Books, Inc., 1959); Joseph Agassi, "Science in Flux: Footnotes to Popper," in *Boston Studies in the Philosophy of Science*, Vol. 3, ed. R. S. Cohen and M. W. Wartofsky (New York: Humanities Press, Inc., 1968), pp. 293–323; I. C. Jarvie, "Nadel on the Aims and Methods of Social Anthropology," *British Journal for the Philosophy of Science*, 12 (1961):1–24; Imre Lakatos, "Falsification and the Methodology of Scientific Research Programmes," in *Criticism and the Growth of Knowledge*, ed. I. Lakatos and A. Musgrave (New York: Cambridge University Press, 1970), pp. 91–196.

17. For example, see Carl G. Hempel, *Philosophy of Natural Science*, Ch. 4.

18. See Martin Gardner, *Fads and Fallacies in the Name of Science* (New York: Dover Publications, Inc., 1957), Ch. 19.

19. For further discussion see Marguerite Foster and Michael Martin, eds., *Probability, Confirmation and Simplicity* (New York: Odyssey Press, 1966) and Day, *Inductive Probability.*

20. For the relevant passages from Hume see M. Foster and M.

Martin, *Probability, Confirmation and Simplicity*, pp. 341–342.

21. Various approaches to this problem are presented in Foster and Martin, *Probability, Confirmation and Simplicity*, Sec. 4.

22. See Popper, *The Logic of Scientific Discovery.*

23. For this critique see Nelson Goodman, "Safety, Strength, and Simplicity," *Philosophy of Science,* 28 (1961):150–151.

24. See Michael Martin, "The Falsifiability of Curve-Hypotheses," *Philosophical Studies,* 16 (1965): 56–60.

25. See Israel Scheffler, *Anatomy of Inquiry*, p. 146.

26. Biological Science Curriculum Study, p. 25.

27. *Ibid.,* p. 26.

28. Earth Science Curriculum Project, *Investigating the Earth* (Boston: Houghton Mifflin Company, 1967), p. 61.

29. Biological Science Curriculum Study, p. 625.

30. *Ibid.,* p. 37.

31. See Martin Gardner, *Fads and Fallacies in the Name of Science,* Ch. 1.

32. See Joseph Schwab, "Structure of the Disciplines: Meanings and Significances," p. 14.

33. For a discussion of this method see Leopold Klopfer, "The Use of Case Histories in Science Teaching," in *Readings in Science Education for the Secondary School,* pp. 226–233.

34. This research paper approach is usually associated with Joseph Schwab, "The Teaching of Science as Enquiry," pp. 73–79. He does not, of course, consider studying pseudoscience "research" papers. See also Howard Baumel and J. Joel Berger, "Teaching from Research Papers: An Approach to Teaching Science as a Process," in *Readings*

in *Science Education for the Secondary School,* pp. 205–207.

35. See Martin Gardner, *Fads and Fallacies in the Name of Science,* Ch. 10.

36. See Michael Martin, "The Use of Pseudo-Science in Science Education," *Science Education,* 55 (1971):53–56.

37. See, for example, James B. Conant, *Science and Common Sense* (New Haven, Conn.: Yale University Press, 1961), pp. 45, 321; and Merritt Kimball, "Understanding the Nature of Science: A Comparison of Scientists and Science Teachers," p. 111.

38. This distinction is lucidly made by Richard Rudner, *Philosophy of Social Science* (Englewood Cliffs, N.J.: Prentice-Hall, Inc., 1966), pp. 4–9. We shall show in Chapter Five how this confusion may affect the validity of *The Nature of Science Scale.*

CHAPTER TWO

1. This strand of the ambiguity is carefully analyzed by Jane R. Martin, *Explaining, Understanding, and Teaching* (New York: McGraw-Hill Book Co., 1970), Ch. 2.

2. Karl Popper, *The Logic of Scientific Discovery.*

3. Carl G. Hempel, *Aspects of Scientific Explanation* (New York: The Free Press, 1965).

4. Ernest Nagel, *The Structure of Science* (New York: Harcourt Brace Jovanovich, Inc., 1961).

5. For a typical criticism see Michael Scriven, "Explanations, Predictions, and Laws," in *Minnesota Studies in the Philosophy of Science,* vol. 3, ed. H. Feigl and G. Maxwell (Minneapolis: University of Minnesota Press, 1962), pp. 170–230.

6. For a defense see May Brodbeck, "Explanation, Prediction, and 'Imperfect' Knowledge," in

Minnesota Studies in the Philosophy of Science, vol. 3, pp. 231–272.

7. Hempel, *Aspects of Scientific Explanation*, pp. 423–425.

8. *Ibid.*, pp. 381–403.

9. I am indebted here to the discussion of Hugh Lehman, "On the Form of Explanation in Evolutionary Theory," *Theoria*, 32 (1966):14–24.

10. Hempel, *Aspects of Scientific Explanation*, p. 352.

11. For a discussion of this point see Scheffler, *Anatomy of Inquiry*, pp. 53–57.

12. *Ibid.*, pp. 46–57.

13. See S. Bromberger, "The Concept of Explanation," Ph.D. dissertation, Harvard University, 1960.

14. Biological Science Curriculum Study, pp. 381–382.

15. Earth Science Curriculum Project, p. 428.

16. For an extended discussion of this and related points see J. Martin, *Explaining, Understanding, and Teaching*, Ch. 9.

17. For a logical analysis of this case see Edward H. Madden, ed., *The Structure of Scientific Thought* (Boston: Houghton Mifflin Company, 1960), pp. 3–7; and Hempel, *Philosophy of Natural Science*, Ch. 1.

18. Hempel, *Philosophy of Natural Science*, Ch. 1 & 2, lucidly analyzes this case.

19. See the analysis in Hempel, *Aspects of Scientific Explanation*, pp. 415–418.

20. See Ernest Nagel, "Teleological Explanation and Teleological Systems," in *Readings in the Philosophy of Science*, pp. 537–558. I am indebted to Hugh Lehman, "Functional Explanation in Biology," *Philosophy of Science*, 32 (1965):1–19, for my analysis.

21. Lehman, "Functional Explanation in Biology," p. 16.

22. Hempel, *Aspects of Scientific Explanation*, pp. 297–330.

23. Lehman, "Functional Explanation in Biology," p. 19.

24. Jerome Bruner, *The Process of Education* (Cambridge, Mass.: Harvard University Press, 1961), p. 28.

25. David P. Ausubel, "An Evaluation of the B.S.C.S. Approach to High School Biology," in *Readings in Science Education for the Secondary School*, p. 387.

CHAPTER THREE

1. For a different way of construing definition see Richard Robinson, *Definition* (London: Oxford University Press, 1950), p. 27.

2. See Stephen F. Barker, *The Elements of Logic* (New York: McGraw-Hill Book Co., 1965), pp. 198–199.

3. The term "explication" was perhaps first clearly and explicitly understood in this way by Carnap, *Logical Foundations of Probability*, Ch. 1.

4. Cf. Barker, *The Elements of Logic*, p. 203.

5. Cf. Hempel, *Philosophy of Natural Science*, Ch. 7.

6. The notion of semantic relevance as well as some of my examples are borrowed from Peter Achinstein, *Concepts of Science* (Baltimore: The Johns Hopkins Press, 1968).

7. Willard V. O. Quine, "Two Dogmas of Empiricism," in *From a Logical Point of View* (Cambridge, Mass.: Harvard University Press, 1953), pp. 20–46; and Morton White, "The Analytic and the Synthetic: An Untenable Dualism," in *Semantics and the Philosophy of Language*, ed. L. Linsky (Urbana: University of Illinois Press, 1952), pp. 272–286. The notion of semantic relevance is linked by

Achinstein, *Concepts of Science*, pp. 39–40, directly to the traditional notion of analyticity. Achinstein does not attempt to answer the major objections raised by Quine and White to analyticity and kindred notions. For example, Achinstein uncritically uses the notion of a logically necessary property. According to Achinstein, some semantically relevant properties are logically necessary for the application of a term. Now this logical necessity is clearly not an extensional notion, but is based on considerations of the connotation or sense of the terms at issue. The old question of how one is to understand this sort of logical necessity remains unanswered. Furthermore, Achinstein's linking of semantic relevance to analyticity has problems of its own. Since a property can be semantically relevant without being logically necessary, an analytic statement can presumably be false. Thus the sentence "Copper conducts electricity" could be used to express an analytic statement, since the property of conducting electricity is semantically relevant for the application of the term "copper." Nevertheless, since conducting electricity is not logically necessary for the application of the term "copper," the statement expressed by the sentence could be analytic *and* false since metal could be copper and *not* conduct electricity. This is not in keeping with what was traditionally understood by analytic statements. Traditionally, analytic statements were true and necessarily so.

8. On this point see Israel Scheffler, *Science and Subjectivity* (Indianapolis: The Bobbs-Merrill Co., Inc., 1967), Ch. 3.

9. See the discussion of extensional isomorphism in Scheffler, *Anatomy of Inquiry*, p. 157.

10. For this argument see Thomas Kuhn, *The Structure of Scientific Revolutions* (Chicago: The University of Chicago Press, 1962); and P. K. Feyerabend, "Explanation, Reduction, and Empiricism," in *Minnesota Studies in the Philosophy of Science*, vol. 3, pp. 28–97.

11. This argument was first explicitly formulated by Scheffler, *Science and Subjectivity*, Ch. 3.

12. This argument is developed at greater length in M. Martin, "Referential Variance and Scientific Objectivity," *British Journal for the Philosophy of Science*, 22 (1971):17–26.

13. Biological Science Curriculum Study, p. 19.

14. B. D. Van Evera, "Definitions, Didactics, and Deliberations," *The Science Teacher*, 25 (April 1958):126.

15. For a detailed analysis of Bridgman's operationism see A. Cornelius Benjamin, *Operationism* (Springfield, Ill.: Charles C Thomas, Publisher, 1955).

16. P. W. Bridgman, *Logic of Modern Physics* (New York: The Macmillan Co., 1927).

17. For further critical discussion of operationism see Hempel, *Philosophy of Natural Science*, Ch. 7, and Michael Martin, "An Examination of the Operationists' Critique of Psychoanalysis," *Social Science Information*, 8 (1969):65–85.

18. W. Mays, "The Epistemology of Professor Piaget," *Aristotelian Society Proceedings*, 54 (1953–1954):57–58.

19. Bruner, *The Process of Education*, p. 26.

CHAPTER FOUR

1. Achinstein, *Concepts of Science*, p. 161.

2. See Fred Dretske, *Seeing and Knowing* (London: Routledge and Kegan Paul, 1969), Ch. 6.
3. Cf. *ibid.*, Ch. 3.
4. Cf. *ibid.*, Ch. 3 & 4.
5. Jerome Bruner and Leo Postman, "On the Perception of Incongruity: A Paradigm," *Journal of Personality*, 18 (1949): 206–223.
6. See Kuhn, *The Structure of Scientific Revolutions*, Ch. 6.
7. Foster and Martin, *Probability, Confirmation and Simplicity*, p. 5.
8. Some Popperians seem to suppose this. See, for example, Jarvie, "Nadel on the Aims and Methods of Social Anthropology," p. 19.
9. See Kuhn, *The Structure of Scientific Revolutions*, Ch. 10.
10. *Ibid.*, p. 112.
11. Cf. Hempel, *Aspects of Scientific Explanation*, pp. 433–447.
12. Day, *Inductive Probability*, p. 275.
13. Israel Scheffler, *Conditions of Knowledge* (Glenview, Ill.: Scott Foresman and Company, 1965), pp. 36–39.
14. Cf. Scheffler, *Science and Subjectivity*, pp. 116–124.
15. Cf. *ibid.*, Ch. 1 & 2.
16. Kuhn, *The Structure of Scientific Revolutions*, pp. 125–126, 144–145; see also Feyerabend, "Explanation, Reduction, and Empiricism."
17. See Norwood Russell Hanson, *Patterns of Discovery* (New York: Cambridge University Press, 1958), for further discussion of theory-laden observation.
18. Scheffler, *Science and Subjectivity*.
19. Cf. Achinstein, *Concepts of Science*, Ch. 5.
20. Earth Science Curriculum Project, p. 3.
21. *Ibid.*, p. 64.
22. M. Martin, "Anomaly Recognition and Research in Science Education," *Journal of Research in Science Teaching*, 7 (1971): 187–190.
23. Scheffler, *Science and Subjectivity*, pp. 86–87.
24. The basic idea of proliferation of theories I derive from Feyerabend, "Explanation, Reduction, and Empiricism."
25. For further details see M. Martin, "Understanding and Participant Observation in Cultural and Social Anthropology," *Boston Studies in the Philosophy of Science*, 4, ed. R. S. Cohen and M. W. Wartofsky (Dordrecht, Holland: D. Reidel Publisher Co., 1969), pp. 303–330.

CHAPTER FIVE
1. This kind of knowledge is sometimes called *knowledge how*. See Gilbert Ryle, *The Concept of Mind* (New York: Barnes & Noble, Inc., 1949), Ch. 2.
2. Scheffler, *Conditions of Knowledge*.
3. Benjamin S. Bloom *et al.*, *Taxonomy of Educational Objectives* (New York: David McKay Co., Inc., 1956), p. 28.
4. For a discussion of stipulative definitions see Chapter Three above.
5. Cf. Scheffler, *Conditions of Knowledge*, pp. 66–74.
6. See *ibid.*, Ch. 5. See also Ryle, *The Concept of Mind*.
7. For an illuminating analysis of understanding see J. Martin, *Explaining, Understanding, and Teaching*, Ch. 7 & 8.
8. Educational Policies Commission, *Education and the Spirit of Science* (Washington, D.C.: National Education Association, 1966), p. 15.
9. *Ibid.*, p. 18.
10. *Ibid.*, p. 19.
11. *Ibid.*, p. 17.
12. David Hawkins, "Education and the Spirit of Science: Critique of a Statement," in *Readings*

in Science Education for the Secondary School, p. 26.

13. Cf. Fletcher G. Watson, "Basic Difficulties in Present High School Science Teaching," *Daedalus*, 1959, 187–191.

14. Cf. James B. Conant, *On Understanding Science* (New York: The New American Library, Inc., 1953).

15. J. Myron Atkin, "Behavioral Objectives in Curriculum Design: A Cautionary Note," in *Readings in the Philosophy of Education: A Study of Curriculum*, ed. J. Martin (Boston: Allyn & Bacon, Inc., 1970), p. 32.

16. Will French, *Behavioral Goals of General Education in High School* (New York: Russell Sage Foundation, 1957), p. 101.

17. Kurtz, "Help Stamp Out Non-Behavioral Objectives," p. 143.

18. French, *Behavioral Goals of General Education in High School*, p. 100.

19. Kurtz, "Help Stamp Out Non-Behavioral Objectives," p. 143.

20. *Ibid.*

21. See Kimball, "Understanding the Nature of Science: A Comparison of Scientists and Science Teachers."

22. For a critical discussion of these views see Chapters One and Three above.

23. Cf. Atkin, "Behavioral Objectives in Curriculum Design: A Cautionary Note."

24. The above view has some similarities to a well-known view in ethics. See, for example, Richard B. Brandt, *Ethical Theory* (Englewood Cliffs, N.J.: Prentice-Hall, Inc., 1959), Ch. 10.

25. This dialectic justification would be a means of justifying the *value* of the propensities suggested by the Commission. It would not by itself show that these propensities should be a goal of *science* education. For this justification the instrumentalist and naturalist approaches would have to be used.

26. Education Policy Commission, p. 16.

27. I am indebted here to the illuminating paper of Joanne Reynolds Bronars, "Tampering with Nature in Elementary School Science," in *Readings in the Philosophy of Education: A Study of Curriculum*, pp. 274–279.

BIBLIOGRAPHY

Achinstein, Peter. *Concepts of Science.* Baltimore, The Johns Hopkins Press, 1968.

Agassi, Joseph. "Science in Flux: Footnotes to Popper." *Boston Studies in the Philosophy of Science,* vol. 3, ed. R. S. Cohen and M. W. Wartofsky. New York, Humanities Press, Inc., 1968. Pp. 293–323.

Andersen, Hans O., ed. *Readings in Science Education for the Secondary School.* New York, The Macmillan Co., 1969.

Atkin, J. Myron. "Behavioral Objectives in Curriculum Design: A Cautionary Note." In *Readings in the Philosophy of Education: A Study of Curriculum,* ed. J. Martin. Boston, Allyn & Bacon, Inc., 1970. Pp. 32–38.

Ausubel, David P. "An Evaluation of the B.S.C.S. Approach to High School Biology." In *Readings in Science Education for the Secondary School,* ed. Hans O. Andersen. New York, The Macmillan Co., 1969. Pp. 374–389.

——. "Some Psychological and Educational Limitations of Learning by Discovery." In *Readings in Science Education for the Secondary School,* ed. Hans O. Andersen. New York, The Macmillan Co., 1969. Pp. 97–114.

Barker, Stephen F. *The Elements of Logic.* New York, McGraw-Hill Book Co., 1965.

Baumel, Howard, and J. Joel Berger. "Teaching from Research Papers: An Approach to Teaching Science as a Process." In *Readings in Science Education for the Secondary School,* ed. Hans O. An-

dersen. New York, The Macmillan Co., 1969. Pp. 205–207.

Benjamin, A. Cornelius. *Operationism.* Springfield, Ill., Charles C Thomas, Publisher, 1955.

Berlin, Barney, and Alan Gaines. "Use Philosophy to Explain the Scientific Method." *The Science Teacher,* 33 (May 1966), 52.

Biological Science Curriculum Study. *Biological Science: An Inquiry into Life.* New York, Harcourt Brace Jovanovich, Inc., 1968.

Bloom, Benjamin S., et al. *Taxonomy of Educational Objectives.* New York, David McKay Co., Inc., 1956.

Brandt, Richard B. *Ethical Theory.* Englewood Cliffs, N.J., Prentice-Hall, Inc., 1959.

Bridgman, P. W. *Logic of Modern Physics.* New York, The Macmillan Co., 1927.

Brodbeck, May. "Explanation, Prediction, and 'Imperfect' Knowledge." In *Minnesota Studies in the Philosophy of Science,* vol. 3, ed. H. Feigl and G. Maxwell. Minneapolis, University of Minnesota Press, 1962. Pp. 231–272.

——. "The Nature and Function of the Philosophy of Science." In *Readings in the Philosophy of Science,* ed. H. Feigl and M. Brodbeck. New York, Appleton-Century-Crofts, 1953. Pp. 3–7.

Bromberger, S. "The Concept of Explanation." Ph.D. thesis, Harvard University, 1960.

Bronars, Joanne Reynolds. "Tampering with Nature in Elementary School Science." In *Readings in the Philosophy of Education: A Study of Curriculum,* ed. J. Martin. Boston, Allyn & Bacon, Inc., 1970. Pp. 274–279.

Bruner, Jerome. *The Process of*

Education. Cambridge, Mass., Harvard University Press, 1961.

———, **and Leo Postman.** "On the Perception of Incongruity: A Paradigm." *Journal of Personality*, 18 (1949), 206–223.

Carnap, Rudolf. *Logical Foundations of Probability.* Chicago, The University of Chicago Press, 1962.

Conant, James B. *On Understanding Science.* New York, The New American Library, Inc., 1953.

———. *Science and Common Sense.* New Haven, Conn., Yale University Press, 1961.

Day, John Patrick. *Inductive Probability.* New York, Humanities Press, Inc., 1961.

Dretske, Fred. *Seeing and Knowing.* London, Routledge and Kegan Paul, 1969.

Earth Science Curriculum Project. *Investigating the Earth.* Boston, Houghton Mifflin Company, 1967.

Educational Policies Commission. *Education and the Spirit of Science.* Washington, D.C., National Education Association, 1966.

Farre, G. L. "On the Problem of Scientific Discovery." *The Science Teacher*, 33 (October 1966), 26–29.

Feyerabend, P. K. "Explanation, Reduction, and Empiricism." *Minnesota Studies in the Philosophy of Science*, vol. 3, ed. H. Feigl and G. Maxwell. Minneapolis, University of Minnesota Press, 1962. Pp. 28–97.

Foster, Marguerite, and Michael Martin, eds. *Probability, Confirmation and Simplicity.* New York, Odyssey Press, 1966.

Frankena, William. *Philosophy of Education.* New York, The Macmillan Co., 1965.

French, Will. *Behavioral Goals of General Education in High School.* New York: Russell Sage Foundation, 1957.

Gardner, Martin. *Fads and Fallacies in the Name of Science.* New York, Dover Publications, Inc., 1957.

Goodman, Nelson. "Safety, Strength, and Simplicity." *Philosophy of Science*, 28 (1961), 150–151.

Hanson, Norwood Russell. "Is There a Logic of Discovery?" In *Current Issues in the Philosophy of Science*, ed. H. Feigl and G. Maxwell. New York, Holt, Rinehart & Winston, Inc., 1961. Pp. 20–35.

———. *Patterns of Discovery.* New York, Cambridge University Press, 1958.

Hawkins, David. "Education and the Spirit of Science: Critique of a Statement." In *Readings in Science Education for the Secondary School*, ed. Hans O. Andersen. New York, The Macmillan Co., 1969. Pp. 25–28.

Hempel, Carl G. *Aspects of Scientific Explanation.* New York, The Free Press, 1965.

———. *Philosophy of Natural Science.* Englewood Cliffs, N.J., Prentice-Hall, Inc., 1966.

Jarvie, I. C. "Nadel on the Aims and Methods of Social Anthropology." *British Journal for the Philosophy of Science*, 12 (1961), 1–24.

———. *The Revolution in Anthropology.* New York: Humanities Press, Inc., 1964.

Kimball, Merritt. "Understanding the Nature of Science: A Comparison of Scientists and Science Teachers." *Journal of Research in Science Teaching*, 5 (1967–1968), 110–120.

Klopfer, Leopold. "The Use of Case Histories in Science Teaching." In *Readings in Science Edu-*

cation for the Secondary School, ed. Hans O. Andersen. New York, The Macmillan Co., 1969. Pp. 226–233.

Kuhn, Thomas. *The Structure of Scientific Revolutions.* Chicago, The University of Chicago Press, 1962.

Kurtz, Edwin. "Help Stamp Out Non-Behavioral Objectives." In *Readings in Science Education in the Secondary School*, ed. Hans O. Andersen. New York, The Macmillan Co., 1969. Pp. 142–145.

Lakatos, Imre. "Falsification and the Methodology of Scientific Research Programmes." In *Criticism and the Growth of Knowledge*, ed. I. Lakatos and A. Musgrave. New York, Cambridge University Press, 1970. Pp. 91–196.

Lehman, Hugh. "Functional Explanation in Biology." *Philosophy of Science*, 32 (1965), 1–19.

————. "On the Form of Explanation in Evolutionary Theory." *Theoria*, 32 (1966), 14–24.

Madden, Edward H., ed. *The Structure of Scientific Thought.* Boston, Houghton Mifflin Company, 1960.

Martin, Jane R. *Explaining, Understanding, and Teaching.* New York, McGraw-Hill Book Co., 1970.

————, ed. *Readings in the Philosophy of Education: A Study of Curriculum.* Boston, Allyn & Bacon, Inc., 1970.

Martin, Michael. "An Examination of the Operationists' Critique of Psychoanalysis." *Social Science Information*, 8 (1969), 65–85.

————. "Anomaly Recognition and Research in Science Education." *Journal of Research in Science Teaching*, 7 (1971), 187–190.

————. "Referential Variance and Scientific Objectivity." *British Journal for the Philosophy of Science*, 22 (1971), 17–26.

————. "The Falsifiability of Curve-Hypotheses." *Philosophical Studies*, 16 (1965), 56–60.

————. "The Use of Pseudo-Science in Science Education." *Science Education*, 55 (1971), 53–56.

————. "Understanding and Participant Observation in Cultural and Social Anthropology." *Boston Studies in the Philosophy of Science*, vol. 4, ed. R. S. Cohen and M. W. Wartofsky. Dordrecht, Holland, D. Reidel Publishing Company, 1969. Pp. 303–330.

May, W. "The Epistemology of Professor Piaget." *Aristotelian Society Proceedings*, 54 (1953–1954), 49–76.

Micciche, Frank, and Michael Keany. "Hypothesis Machine." *The Science Teacher*, 36 (April 1969), 53–54.

Nagel, Ernest. "Principles of the Theory of Probability." *International Encyclopedia of Unified Science*, 1, no. 6 (1939).

————. "Teleological Explanation and Teleological Systems." In *Readings in the Philosophy of Science*, ed. H. Feigl and M. Brodbeck. New York, Appleton-Century-Crofts, 1953. Pp. 537–558.

————. *The Structure of Science.* New York, Harcourt Brace Jovanovich, Inc., 1961.

Pap, Arthur. *An Introduction to the Philosophy of Science.* New York, The Free Press, 1962.

Popper, Karl. *The Logic of Scientific Discovery.* New York, Basic Books, Inc., 1959.

Quine, Willard V. O. "Two Dogmas of Empiricism." In *From a Logical Point of View.* Cambridge, Mass., Harvard University

Press, 1953. Pp. 20–46.

Reichenbach, Hans. *Experience and Prediction.* Chicago, The University of Chicago Press, 1938.

Robinson, James T. "Philosophical and Historical Bases of Science Teaching." *Review of Educational Research,* 39 (1969), 459–471.

Robinson, Richard. *Definition.* London, Oxford University Press, 1950.

Rudner, Richard. *Philosophy of Social Science.* Englewood Cliffs, N. J., Prentice-Hall, Inc., 1966.

Rutherford, F. James. "The Role of Inquiry in Science Teaching," *Journal of Research in Science Teaching,* 2 (1964), 80–84.

Ryle, Gilbert. *The Concept of Mind.* New York, Barnes & Noble, Inc., 1949.

Salmon, Wesley C. *The Foundations of Scientific Inference.* Pittsburgh, University of Pittsburgh Press, 1966.

Scheffler, Israel. *Conditions of Knowledge.* Glenview, Ill., Scott, Foresman and Company, 1965.

———. *Science and Subjectivity.* Indianapolis, The Bobbs-Merrill Co., Inc., 1967.

———. *Anatomy of Inquiry.* New York, Alfred A. Knopf, Inc., 1963.

Schwab, Joseph. "Structure of the Disciplines: Meanings and Significances." In *The Structure of Knowledge and the Curriculum,* ed. G. W. Ford and Lawrence Pugno. Chicago, Rand McNally and Co., 1964. Pp. 1–30.

———. "The Teaching of Science as Enquiry." In *The Teaching of Science.* Cambridge, Mass., Harvard University Press, 1964. Pp. 1–104.

Scriven, Michael. "Explanations, Predictions, and Laws." *Minnesota Studies in the Philosophy of Science,* vol. 3, ed. H. Feigl and G. Maxwell. Minneapolis, University of Minnesota Press, 1962. Pp. 170–230.

Shoresman, Peter. "A Technique to Clarify the Nature of Theories." *The Science Teacher,* 32 (May 1965), 53–55.

Van Evera, B. D. "Definitions, Didactics, and Deliberations." *The Science Teacher,* 25 (April 1958), 125–126, 154–157.

Wallas, Graham. *The Art of Thought.* New York, Harcourt Brace Jovanovich, Inc., 1926.

Watson, Fletcher G. "Basic Difficulties in Present High School Science Teaching." *Daedalus,* 1959, pp. 187–191.

White, Morton. "The Analytic and the Synthetic: An Untenable Dualism." In *Semantics and the Philosophy of Language,* ed. L. Linsky. Urbana, University of Illinois Press, 1952. Pp. 272–286.

INDEX

151–152; of deductive-nomological explanation, 51–52; of nondemonstrative inference, 29–31; in propositional knowledge, 136–137. *See also* Evidence.

K

Kekulé von Stradonitz, Friedrich: on hypothesis generation, 10,12
Kepler, Johannes: and logic of discovery, 15; and Newton's laws, 58–59; and third law of planetary motion, 25, 27
Knowledge: forms of, 142; as goal of science education, 133–137, 138–140, 146; propensities for, 140, 143–144
Kuhn, Thomas S.: on observation, 116
Kurtz, Edwin: on behavioral goals, 148, 149, 150

L

Language: mentalistic, 149; in observation, 121–122; in science teaching, 126–127
Laws. *See* Scientific laws.
Lehman, Hugh: on functional explanations, 72
Lever law, 25; test implications of, 26
Life: definition of, 89–90
Logic: of confirmation approach, 18; of definition, 82, 87–88; of discovery, 14–15; of explanation, 48–49, 51–53, 59; of nondemonstrative inference, 30; and Redi, 37; of refutation approach, 35; in science education, 67, 68; in spontaneous generation theory, 19–23; in testing of premises, 143
Logical positivist, 34
Logic of Modern Physics, The, 95

M

Magnetism: operational definition of, 96–97, 98
Mays, W.: on operationism, 100–101
Mentalistic concepts: testability of, 152; use of, 149, 150
Methodological rules (MR): in operational definitions, 95–100
Morality: and science education, 158–160
Motivation: for knowledge, 142, 143–144; in observation, 111–112. *See also* Propensities.
MR. *See* Methodological rules.

N

Nagel, Ernest, 50
Natural science: experimentation in, 25
Naturalistic justification, 155–157
Newton, Sir Isaac: force defined by, 78–79; and Kepler's law, 27, 58–59; theory of, 26
Noncausal explanations, 58–60
Noncognitive observation, 105, 107, 128, 129
Nondemonstrative inference: justification for, 29–31; types of, 28–29. *See also* Generalization.
Nonextensional contexts, 86
Nonscience: refutation approach to, 34–35, 36

O

Objectivity in science: and definition, 86–88; and language, 114; and observation, 116–121
Observation, 103–131, 142; and behavioral goals, 148; methods of, 43; and naturalistic justification, 155; science educators' view of, 152
Operational definitions, 94–100
Operationism: in physics, 94–95
Ostensive teaching of words, 76, 91, 101

P

Participant observation, 128–131
Pedagogical rules (PR): in operationism, 101–102; in science education goals, 153–154
Pendulum: noncausal explanation of, 59
Phenomenological immediacy, 105, 106
Physics: observational language in, 122; operationism in, 94–100
Piaget, Jean: and operationism, 100–101
Planetary motion: third law of, 25, 27
Pluto: and test of Kepler's law, 25
Popper, Karl: and deductive-nomological explanation, 50; and refutation approach, 18, 32–36, 39–40
Postman, Leo: and experiments in perception, 108; and premise influence, 119, 120
PR. *See* Pedagogical rules.
Prediction, 62–64, 144–145
Predictive inferences, 29, 30, 31
Premise influence: on observation, 119–120; prevention of, 124
Preparation stage: of hypothesis generation, 11–12, 13, 16–17
Primary cognitive observing, 105, 107

Teaching of science. *See* Science teachers.

Test implications: in behavioral goals, 153–154; in operationism, 99–100; in scientific inquiry, 22, 23, 24–25

Testing, 6, 17–36; observation in, 109, 111, 114, 117, 119, 122, 129; operational definitions in, 98–100; propensity for, 141–143; and science education, 36–43, 92, 102, 147–157

Theoretical science: observation in, 107–108; prediction in, 63, 145. *See also* Theory.

Theory: generation of, 142; objectivity of, 86–88, 116–120, 124–126; observation influenced by, 106–116; in operationism, 99–100; prediction from, 145; and scientific understanding, 58–59, 61, 62. *See also* Hypotheses.

Theory boxes, 16–17

Truth: in deductive-nomological explanation, 51; in definitions, 82–83, 85–86; in hypothesis testing, 17–18; and premise influence, 124–125; and propositional knowledge, 134, 135. *See also* Knowledge, Scientific laws.

U

Understanding. *See* Scientific understanding.

V

Visual perception. *See* Observation.

W

Wallas, Graham: and stages of hypothesis generation, 11–12, 13, 14

"Working hypotheses": in observation, 124–126